JOHN VERHOOGEN
University of California,
Berkeley

Energetics of the Earth

NATIONAL ACADEMY OF SCIENCES
Washington, D.C.   1980

The National Academy of Sciences was established in 1863 by Act of Congress as a private, non-profit, self-governing membership corporation for the furtherance of science and technology, required to advise the federal government upon request within its fields of competence. Under its corporate charter the Academy established the National Research Council in 1916, the National Academy of Engineering in 1964, and the Institute of Medicine in 1970.

Library of Congress Cataloging in Publication Data

Verhoogen, John, 1912-
    Energetics of the Earth.

    Bibliography: p.
    Includes index.
    1. Earth temperature. 2. Earth—Internal structure.
I. Title.
QE509.V45   551.1'2   80-17501
ISBN 0-309-03076-5

Available from

National Academy Press
National Academy of Sciences
2101 Constitution Avenue, N.W.
Washington, D.C. 20418

Printed in the United States of America

Preface

This is the text, somewhat enlarged, of a series of four lectures delivered at Stanford University in January 1979 by the happy recipient of the 1978 Arthur L. Day Prize and Lectureship of the National Academy of Sciences. Funds are earmarked for publication of the lectures, which according to the terms of the award should "prove a solid, timely and useful addition to the knowledge and literature in the field."

Whether the lectures will prove to be solid, timely, or useful is as uncertain as their topic. This topic, which has been of great interest to me for more than 40 years, remains largely speculative. Progress has been slow; there is not so terribly much more to say about temperature at the core-mantle boundary than Francis Birch said in 1952 in his celebrated paper on the constitution of the earth's interior. As the reader will readily perceive, there is still room for considerable difference of opinion on almost every subject.

Much of the uncertainty stems from our ignorance of physical properties of terrestrial materials at pressures of a few megabars, such as exist in the lower mantle and core. One may hope that much more information will become available in the next few years, now that static experiments appear feasible in the megabar range. On the other hand, our understanding of thermal events at the time of, or shortly after, formation of the earth will remain speculative for many years to come, as also will the crucial matter of the abundance of radioactive elements in the mantle and core. On the theoretical side, perhaps the greatest step forward in the last four decades—or indeed since the discovery of radioactivity—has been the recognition of convection as the dominant mode of heat transfer in most of the mantle and core. Here, at long last, is something to hang on to.

The writer wishes to express deep thanks to all his colleagues at Berkeley who have helped him throughout the years, and particularly to the graduate students who have discussed these matters with him in seminars. He is grateful also to the audience at Stanford University for their warm welcome, and to Professor Allan Cox for the perfection of the arrangements.

Berkeley, March 1979

Contents

1 ENERGY SINKS   1

  Heat Flow, 2
  Volcanic Heat, 4
  Metamorphic Heat, 5
  Strain Energy; Earthquakes, 7
  Potential Energy; Uplift of Mountains, 8
  Plate Tectonics, 8
  Kinetic Energy of Rotation, 11
  Summary, 13

2 ENERGY SOURCES   15

  Original Heat, 15
  Gravitational Energy, 18
      Core Formation, 19
  Tidal Friction, 21
  Radiogenic Heat, 23
      Crustal Heat Versus Mantle Heat, 26
  Summary, 28

3 TEMPERATURES WITHIN THE EARTH   29

  The Upper Mantle, 31
      The Low-Velocity Zone, 31;
      Geotherms from Nodules in Kimberlite, 33

The Mantle's Transition Zones, 35
The Lower Mantle, 36
  Layer D', 36; Layer D", 44
The Outer Core, 49
  Composition of the Core, 49; Temperature from an Equation of State, 52; The Adiabatic Gradient, 54
The Temperature at the Inner Core–Outer Core Boundary, 56
  The Melting Point of Iron, 56; Melting in the Fe-S System, 60
Summary, 65

4 DYNAMICS OF THE CORE   67

Stability of the Core, 68
Causes for Convection, 70
Energy Requirements of the Geomagnetic Dynamo, 70
Efficiency of a Steady-State Thermal Dynamo, 75
An Improved Estimate of Efficiency, 78
The Entropy Balance Equation, 83
The Gravitational Dynamo, 88
  Gravitational Energy, 89; Volumetric Relations, 91; Heat Output of Core, 95

5 CORE-MANTLE INTERACTIONS   100

Efficiency of the Mantle, 100
Convection Patterns in the Mantle, 104
  Convective Pattern for Distributed Sources of Heat, 104; Heat from the Core, 110
Additional Factors in Controlling the Flow, 112
  Temperature Dependence of Physical Properties, 112; The Von Zeipel Instability, 112; Effects of Surface Plates, 114; Effects of Phase Transitions in the Mantle, 114
Influence of the Core on Convection in the Mantle, 118
Influence of the Mantle on Convection in the Core, 120
Some Final Remarks, 121

REFERENCES   125

AUTHOR INDEX   133

SUBJECT INDEX   137

Energetics of the Earth

1 Energy Sinks

Every geological process involves exchange or transfer of energy in one form or another. Intrusion of a plutonic body causes heat to be transferred to the wall rock and carried down the temperature gradient, away from the intrusive body; work must be done to strain rocks or to uplift a mountain; heat must be applied to rocks undergoing metamorphism; and so on. Much of this energy is eventually discharged as heat into the oceans and atmosphere, whence it is radiated out into space. Most geological processes thus entail a loss of energy from the earth into what we shall think of as a heat sink.

There is no great difficulty in imagining where this energy could come from. We could, for instance, suppose that the earth was originally endowed with it in the form of the "original heat" of an initially very hot earth. This is essentially the point of view that prevailed throughout the nineteenth century, up to the discovery of radioactivity. We now know that there are other possible sources of energy (e.g., radioactivity, gravitational energy) in addition to the original heat; as we shall see in Chapter 2, there is in fact an almost embarrassing abundance of riches. Our problem is thus a little more subtle. What we want to find out is how the energy gets where we observe it and in its proper form, whether kinetic, potential, chemical, magnetic, or just plain heat.

Let us elaborate for a moment on this last point. Although all forms of energy (e.g., kinetic) are readily transformed into heat, the converse is not true. Heat can never be totally converted to mechanical work; the efficiency of the conversion—that is, the ratio of work output to heat input—depends on, among other things, the temperature differences that must exist for there to be any conversion at all. Regardless of the amount of heat an isolated body may initially contain, very little can happen if the temperature inside it is uniform. To account for geological behavior of the earth, it is necessary to postulate not only adequate heat sources, but also a "structure" of some kind. A structureless, formless, isotropic sea of heat is geologically useless. Much of the inquiry we are now embarking on will center on the manner in which appropriate structures can develop in the spontaneous evolution of a body such as the earth; as we shall see, it is the earth's gravitational field that provides most of the desired structure. Much of the following discussion relates to the interaction of gravitational and thermal fields.

But first we wish to make a brief inventory of the energy requirements of geological (in the broad sense) phenomena, which for present purposes we classify as follows:

1. Heat
    (a) Surface heat flow
    (b) Volcanic heat
    (c) Metamorphic heat
2. Mechanical energy
    (a) Strain energy
    (b) Uplift of mountains
    (c) Motion of plates
    (d) Kinetic energy of rotation
3. Magnetic energy

Item 3, although not very significant in magnitude, will be examined in some detail in Chapter 4 because of its bearing on the nature of the earth's core.

HEAT FLOW

Surface heat flow, or heat flow for short, refers to the rate at which heat flows out across the interface between the solid earth and the atmosphere or oceans. The heat flow $\mathbf{q}$ is determined by measuring the temperature gradient ∇T in near-surface rocks and their thermal conductivity k; by Fourier's law of heat conduction

$$\mathbf{q} = -k\nabla T, \qquad (1.1)$$

where the minus sign reminds us that heat flows spontaneously down the temperature gradient, from a hot point to a colder one. The average of several thousand measurements on land and on the seafloor is about 1.5×10^{-6} cal/cm² s, or 1.5 heat flow units (HFU); 1 HFU is approximately equal to 40 mW/m².

The heat flow determined from (1.1) is a lower bound for its actual value, for in porous and permeable rocks heat may also be carried by convective flow of interstitial fluid (water). This convection lowers the temperature gradient below the value it would have if the rocks were dry or impervious to water. For this reason Williams and von Herzen (1974) have proposed a somewhat higher mean heat flow value of 80 mW/m² ($\approx$2 HFU), corresponding to a heat loss for the whole earth (area = 5.1×10^{14} m²) of about 4×10^{13} W.

Heat flow is observed to vary locally within rather broad limits, so that an arithmetic mean of all observations is not very suitable; ideally, each measurement should be weighted according to the size of the area it represents. The difficulty is that no measurements at all are available in some areas. Chapman and Pollack (1975) have attempted to remedy this lack of data by observing that in many regions heat flow is found to be related to the geologic age of the region (see below); conversely, heat flow values can be predicted in regions where they have not been measured if the geologic age is known. Chapman and Pollack determine in this manner a mean heat flow of 50 mW/m², or 3×10^{12} W for the whole earth; these data, however, are not corrected for convective transport by pore fluids.

A most interesting feature of the heat flow is its regional "structure." It is generally observed on land that local values of heat flow tend to decrease with increasing geologic age of the province, that is, time elapsed since the last magmatic or metamorphic event. Thus the heat flow is usually lower in Archean shield areas ($\approx$40 mW/m²) than in regions of Cenozoic tectonic activity and volcanism ($\approx$80 mW/m²). This regional variation may be related to the uneven vertical distribution in the crust of radioactive heat sources that will be discussed in Chapter 2. Heat flow on oceanic plates also varies with local age, defined (through magnetic lineations) as distance from the ridge axis divided by the plate velocity. Heat flow near a ridge axis may be typically about 100 mW/m², twice its value on portions of a plate that are older than 100 million years. Both the heat flow data and the topography of the ocean floor can be explained (McKenzie, 1967; Sclater and Francheteau, 1970; Parker and Oldenburg, 1973; Chapman and Pollock, 1977) by gradual cooling and contraction of a plate as it moves away from the ridge where it formed by upwelling of hot mantle material. Conversely, the age–heat flow relationship provides rather convincing evidence as to the role of mantle convection in heat transport.

VOLCANIC HEAT

By volcanic heat we mean the heat that is brought to the surface (ocean or atmosphere) by volcanic activity, mainly the outpouring of lava. For each gram of lava that cools from 1000° to 0°C, crystallizing as it cools, about 400 cal ($\approx$1600 J) are released. The average annual rate of outpouring is not exactly known. A single eruption may produce several cubic kilometers of lava and pyroclasts, but such large eruptions are infrequent. From 1952 to 1971, Kilauea volcano on the island of Hawaii produced about 0.11 km^3/yr. Rates of accumulation of plateau basalts seem to be of the same order. It may be surmised that the average rate of eruption on land is less than 1 km^3/yr.

The rate of eruption on the ocean floor is even less well known. Most of the submarine volcanic activity probably occurs in the crestal area of ridges where new oceanic crust is formed. Following Christensen and Salisbury (1975), we assume that the total thickness of lava flows comprising the upper part of the oceanic crust is about 1.5 km. Taking the total length of ridges to be 5×10^4 km and the average plate velocity to be 2.5 cm/yr, the rate of production of crust at a symmetric ridge is 5 cm/yr. The erupted volume of lava then comes out close to 4 km^3/yr. The total volume (land plus ocean) is thus somewhat less than 5 km^3/yr, and the corresponding heat loss is about 8×10^{11} W, a few percent of the heat flow.

This estimate does not include, of course, the heat brought up through the crust, but not quite to the surface, by bodies of magma that cool at depth. This heat will presumably be included in the heat flow. Thus, for instance, the very high heat flow (about 600 mW/m^2, 10 times normal) measured in the Yellowstone caldera beneath Yellowstone Lake (Morgan *et al.*, 1977), which almost certainly comes from a subjacent body of magma, need not be counted as volcanic heat since it is presumably already included in the heat flow data. Similarly, any heat released by cooling of the intrusive gabbros that may form a large part of layer 3 of the oceanic crust is included in the seafloor heat flow measurements.

Thus it would seem that the rate of heat lost to the atmosphere and oceans by surface volcanic activity, on land and on the seafloor, may be about 8×10^{11} W, less than the uncertainty affecting the global heat flow, which, as we have just seen, is variously estimated as 3×10^{13} or 4×10^{13} W. It follows that there should be no great difficulty in finding adequate sources of volcanic energy if we can find adequate sources for the global heat flow. The volcanic problem is again a "structural" one; what is to be explained is why volcanoes are located where they are.

Nothing has been said so far of the energy transferred to the atmosphere through volcanic explosions. A single major explosion, such as occurred at Krakatoa in 1883, might release some 10^{25} ergs, but since such explosions do not occur very frequently (fewer than 10 per century, if that), the total power involved is probably not much greater than 10^9 W, which is negligible.

METAMORPHIC HEAT

By metamorphic heat we mean the heat required to transform sedimentary and volcanic rocks into their recrystallized metamorphic equivalents—greenstones, amphibolites, gneisses, granulites, and so forth. Many metamorphic reactions involve dehydration (e.g., muscovite + quartz → orthoclase + sillimanite + water) or decarbonation, and are endothermic. Thus heat must be applied to heat cold sediments to the reaction temperature and then to make the reaction go. We count the heat of an endothermic reaction as a sink because it effectively remains locked in the metamorphic rock until surface weathering and accompanying hydration and carbonation return a metamorphic rock to its original condition, releasing the heat of reaction into the atmosphere.

When it is possible to reconstruct from the mineralogy of a metamorphic rock the precise conditions of pressure and temperature under which it recrystallized, it rather frequently appears that metamorphic temperatures must have exceeded, perhaps by as much as 100° or 200°, the temperature one would normally expect to prevail at the corresponding pressure or depth. It is difficult to make very precise statements regarding this temperature excess because, in the first place, metamorphic temperatures are not all that easy to determine, as they depend on such things as the difference between the lithostatic pressure to which the solid part of the system is subjected and the pressure of the gas or fluid phase; they may also depend on the composition (e.g., the H_2O/CO_2 ratio) in the fluid phase. It is also difficult to define a "normal" expected temperature since continental geotherms vary according to the surface heat flow and to the vertical distribution of radioactive heat sources. The impression nevertheless remains that some types of high-grade metamorphic rocks, particularly granulites or rocks associated with migmatites and granitic melts, form only when there is an abnormally high rate of heat influx rising into the continental crust from below.

Such surges of heat would probably still be required for endothermic metamorphism even if no "excess" temperature were required. This is seen by referring to Figure 1–1, which represents possible temperature profiles (geotherms) in a homogeneous crust of thickness H to which a constant heat flux q_0 is applied from below. If the crust contains neither sources nor sinks of heat, the steady-state temperature distribution is linear with a slope (gradient) $dT/dz = q_0/k$, where k is the thermal conductivity. If the crust contains heat sources (e.g., radioactivity or exothermic reactions) of intensity $\epsilon > 0$, the steady-state geotherm will be as shown by the upper curve. The gradient at the base of the crust ($z = H$) is the same as before if q_0 is the same, but the gradient at the surface ($z = 0$) is now $(q_0 + \epsilon H)/k$; the surface heat flow equals the heat supplied at the bottom plus the heat generated in the crust itself. If, on the contrary, endothermic reactions are taking place, ϵ is negative, and the steady-state temperature is given by the lower curve, whose slope at the surface is less than that of the two other curves. Note that the effect of the endothermic reactions is to lower the temperature below that of the other curves. Thus endothermic reactions will run at temperatures equal to or greater than the normal temperatures only if the heat flow q_0 is increased by an amount that depends on ϵ, a quantity that is difficult to evaluate because of our total ignorance of the *rate* at

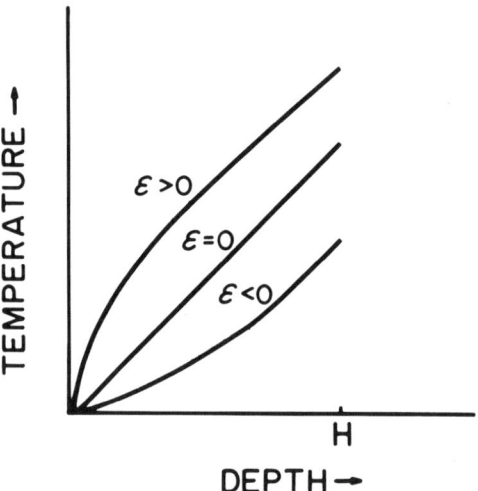

FIGURE 1-1 Effect of exothermic reactions ($\epsilon > 0$) or endothermic reactions ($\epsilon < 0$) on temperature in a plate of thickness H heated from below. The rate of heating from below is the same in all cases, and a steady state is assumed. Temperatures are lower when $\epsilon < 0$ than when $\epsilon \geq 0$.

which metamorphic reactions proceed. As a guideline, a reaction requiring 50 cal/g of rock and running to completion in 10^6 yr is equivalent to a heat sink of 16×10^{-13} cal/g s or roughly 44×10^{-13} cal/cm^3 s, assuming a density of 2.75 g/cm^3. By comparison, radioactive heat generation in granites is only about 6×10^{-13} cal/cm^3 s.

The problem is actually much more complicated than Figure 1-1 suggests. In the first place, it is unlikely that anything like a steady state is ever reached, since the characteristic thermal diffusion time for the crust ($\approx 10^7$ yr) is of the same order as the duration of a metamorphic episode. Secondly, much heat may be carried convectively rather than conductively by fluids (e.g., water released by dehydration). Finally, the assumption of uniform ϵ on which the curves of Figure 1-1 are based is certainly incorrect: radioactivity is not uniformly distributed, and both the heat and the rate of reaction depend critically on temperature. Yet it remains almost certainly true that in a region undergoing regional metamorphism of the high-temperature, low-pressure type, the rate of heat input into the crust at the time of metamorphism may be several times larger than it is now in old continental platforms or shield areas. As deformation and orogeny are commonly associated with regional metamorphism, orogeny should perhaps be described as a thermal disturbance rather than a mechanical one. But since the fraction of the earth's surface undergoing orogeny and regional metamorphism at any time is small, metamorphic heat as defined here is probably but a small fraction of the global heat flow, of the same order perhaps as volcanic heat, and will not be further considered.

STRAIN ENERGY; EARTHQUAKES

In broad terms an earthquake is believed to occur when sudden yielding or fracturing releases strain energy that has slowly accumulated in the neighborhood of the focus. At the time of fracture, some or all of that strain energy E_s is converted to kinetic and potential energy E_r of the radiated seismic waves. E_r can be determined from the amplitude and frequency of these waves, from which a number called the magnitude M of the earthquake is also calculated. The relation between E_r and M is often taken to be

$$\log E_r = 11.3 + 1.8M,$$

where E_r is expressed in ergs and the log is to base 10. From a statistical determination of the average number of earthquakes of given magnitude occurring each year, the average total rate of release of seismic energy is of the order of 10^{26} ergs/yr, or about 3×10^{11} W. This does not represent the whole of E_s, as it does not include the strain energy accumulated in irreversible deformation (e.g., folding) or that part of the reversible strain energy that is not converted to seismic energy (e.g., heat generated by friction along the fault surface). Thus E_s might perhaps be as high as 10^{12} W, roughly 2 or 3 percent of the global heat flux.

POTENTIAL ENERGY; UPLIFT OF MOUNTAINS

To raise a body of sedimentary rocks from below sea level to form a mountain range, work must be done against gravity. Work must also be done to form the root of the mountain by displacing heavier mantle material. The rate at which work is thus converted into potential energy is not easily calculated in the absence of detailed information on the density structure of the crust and mantle prior to and after formation of the mountain; yet a rough average estimate may be obtained if one is willing to assume that the average height of land has remained roughly constant through more recent geologic times. This constancy requires that rates of uplift equal, on the average, rates of erosion. If so, the rate of increase of potential energy must balance, on the average, the rate at which this potential energy is degraded to heat during erosion. A mass m falling through a vertical distance h releases an amount of potential energy gmh, where g is the acceleration of gravity. Let $h = 1$ km be the average height of land, with total area 1.5×10^{18} cm². Let the average rate of erosion (thickness removed in 1 year) be about 5×10^{-3} cm (a rough guess). The volume eroded per year is then $= 7.5 \times 10^{15}$ cm³ and its mass is roughly 2×10^{16} g, assuming a density of 2.5 g/cm³. This mass, falling a vertical distance of 1 km $= 10^5$ cm, will release 2×10^{24} ergs, since $g \cong 10^3$ cm/s². The corresponding input of potential energy to balance this loss is then 2×10^{24} ergs/yr, or 7×10^9 W. This is negligibly small in comparison with the global heat flux of $3-4 \times 10^{13}$ W.

PLATE TECTONICS

Since earthquakes, faulting, and deformation in general are now generally considered to be part of the broader phenomenon of plate motion, it may be appropriate to disregard for the moment the strain energy and look instead at the energy required to drive plates. There seems to be much unnecessary confusion as to what the driving forces are. Some authors have suggested that oceanic plates are pulled by the negative buoyancy of their subducted slabs, which are colder and denser than the surrounding mantle; gravity then pulls them down. Others imagine plates to be driven by the viscous drag of the fluid in the underlying asthenosphere, while still others think of plates as simply sliding down the flanks of oceanic ridges. All three suggestions take only a limited view of the phenomenon.

To illustrate this point, suppose that we ask a proponent of negative buoyancy what causes rain to fall on land. "Well, of course," he will say, "it is negative buoyancy. Liquid water being denser than the surrounding air, gravity must necessarily pull raindrops down." While this is true, it neglects four other essential features of the process, namely (1) the input of solar heat that evaporates water over the ocean, (2) the positive buoyancy of water vapor, which gravity forces to rise into the atmosphere, (3) pressure gradients in the atmosphere that move masses of air from ocean toward land, and (4) cooling that causes water vapor to condense into raindrops. Clearly, without all these factors we would have no rain on land. Plates are moved as much by the positive buoyancy of hot material rising at a ridge as by the negative buoyancy of the downgoing subducted slab, both of which are elements of a single convectional process. Plates move for the same reason the rest of the mantle does. The lithosphere is nothing but a portion of the general flow that has acquired by cooling somewhat different mechanical properties (Figure 1-2). Its thickness, which increases with age or distance from the ridge (Chapman and Pollack, 1977), is a measure of the cooling that has taken place because of heat loss through the oceanic floor.

The forces that determine the flow velocity at any point in a convective system are (1) buoyancy or gravitational forces arising from density differences brought about mainly by temperature differences, (2) a

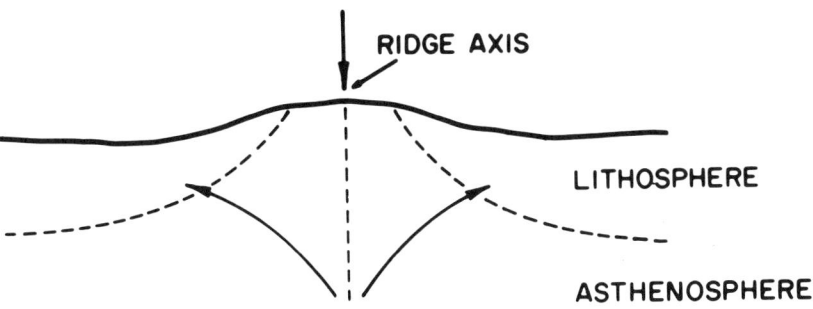

FIGURE 1-2 A spreading ridge. The oceanic lithosphere is part of the mantle; its motion is caused by the same forces that cause the rest of the mantle's flow.

pressure gradient, and (3) viscous forces that can only redistribute momentum and convert kinetic energy to heat and that cannot drive the motion in the absence of other driving mechanisms. It is important to remember that the pressure gradients cannot be calculated by looking at the buoyancy at only one point. The horizontal pressure gradient that drives horizontal flow in the upper level of a conventional Benard convection cell is the result of both the negative buoyancy in the cold descending flow and the positive buoyancy of the ascending flow. It arises essentially from the vertical gravity forces because of the requirement of conservation of matter, which dictates that matter cannot indefinitely accumulate at the top of a rising column, or at the bottom of a sinking one, but must move sideways.

To summarize, the energy that drives plates is the same gravitational energy that drives all the rest of the convective flow. It is measured by the integral $\int_V \rho \mathbf{g} \cdot \mathbf{u}\, dV$, where ρ is density, $\mathbf{g}$ is the acceleration of gravity, $\mathbf{u}$ is velocity, and integration is done over the whole volume of the convecting fluid. For convection to occur, this integral must exceed the sum of the viscous dissipation plus the work done by the system on its surroundings (for instance, the work of deformation done on passive continental plates that are being rafted along). The gravitational integral would be zero if the density were uniform so that div $\mathbf{u} = 0$. Indeed, if ψ is the gravitational potential such that $\mathbf{g} = \nabla \psi$ and div $\mathbf{u} = 0$:

$$\int_V \rho \mathbf{g} \cdot \mathbf{u}\, dV = \rho \int_V \mathbf{u} \cdot \nabla \psi\, dV = \rho \int_V \text{div}\, (\mathbf{u}\psi)\, dV - \rho \int_V \psi\, \text{div}\, \mathbf{u}\, dV$$

$$= \rho \int_S \psi \mathbf{u} \cdot d\mathbf{S} = 0$$

since the normal component of the velocity must vanish on the bounding surface S. In thermal convection, nonuniformity of ρ is maintained by temperature differences that in turn require heat sources to be maintained against the tendency for conduction and convection to equalize both temperature and density. Clearly, for the fluid to expand when heated, work must be done against the prevailing pressure, and this work must come from the heat supply. But in the steady state, exactly the same amount of heat comes out of the system at the top as goes into it at the bottom. The heat that comes out is what we have called the global heat flow, or, more precisely, the global heat flow minus the heat generated in the continental crust. Plate tectonics does not therefore require consideration of an additional energy sink.

As noted above, the continental crust and, to some extent, the thick lithosphere under it must be considered separately, because they are not integral parts of the convective system. Because of its low density, continental crust is not easily subducted, and there is some evidence that very little mixing occurs between continental lithosphere and the rest of the mantle. Continents and their lithosphere both appear to be passively pushed or rafted along on the main mantle flow. Yet work is done on the continental crust, as when mountains form when continents collide. We shall examine briefly in Chapter 5 conditions under which mantle convection can do work on its surroundings.

KINETIC ENERGY OF ROTATION

The kinetic energy of rotation is $E_K = \frac{1}{2}I\omega^2$, where ω is the angular velocity of rotation and I is the moment of inertia about the rotation axis. E_K, which is about 2×10^{29} J, changes as either I or ω changes. In the absence of external torques, the angular momentum $I\omega$ must remain constant. Thus changes in I and ω are related as $I\,d\omega + \omega\,dI = 0$. The kinetic energy change corresponding to a change dI is then

$$dE_K = -\tfrac{1}{2}\omega^2\,dI.$$

Thus, if gravitational separation occurs in the earth, with denser matter moving toward the center, I will decrease and the kinetic energy will increase at the expense of part of the gravitational energy released by the condensation. Conversely, if the earth were to expand because of internal heating, its moment of inertia would increase and its kinetic energy would decrease.

The rate ω is measured by astronomical observations made in observatories fixed on the crust and mantle. What is measured is ω_m, the angular velocity of the mantle, which may differ from ω_c, the rate of rotation of the core. If total angular momentum (mantle + core) remains constant (no external torque), while the moments of inertia of the mantle, I_m, and of the core, I_c, also remain constant, the kinetic energy of the whole earth changes as

$$\frac{dE_K}{dt} = I_m(\omega_m - \omega_c)\frac{d\omega_m}{dt}.$$

Suppose $\omega_m - \omega_c \cong 10^{-10}$/s, corresponding to a mantle that makes one extra turn with respect to the core every 2000 years, a figure suggested by the rate of westward drift of the magnetic secular variation. Changes in the length of the day amounting to a few milliseconds have been observed to occur in the course of a few, say 10, years. The corresponding acceleration $d\omega_m/dt \cong 10^{-20}/\text{s}^2$, and $dE_K/dt \cong 10^8$ W. Since the acceleration of ω_m is presumably caused by an electromagnetic internal torque exerted by the core on the mantle, the corresponding change in kinetic energy must come from the core. Conversely, when the mantle decelerates, energy flows back into the core. The power involved (10^8 W) is negligible in the present context.

Tidal torques of external origin (sun, moon) cause a secular deceleration of the earth of about $5 \times 10^{-22}/\text{s}^2$. Kinetic energy of rotation is dissipated by friction (viscosity of the oceans, anelasticity of the mantle) and reappears as heat (see Chapter 2). Heat generated by viscous dissipation in the oceans is rapidly lost to the earth and must be counted as a sink; heat generated by tidal deformation of the mantle appears there as a source.

An input of energy is also needed to displace the instantaneous axis of rotation away from the principal axis of figure, as happens when the amplitude of the Chandler wobble increases. Excitation of the wobble alternates, however, with episodes of damping, during which the stored kinetic energy is again dissipated as heat in a matter of a few decades. The energy involved is $\frac{1}{2}\omega^2\alpha^2 A(C - A)/C$, where A and C are respectively the minimum and maximum principal moments of inertia of the earth, and α is the small angular separation between C and the rotation axis. Since α is at most about 0.2 arc sec = 10^{-6} rad, the power involved in the Chandler wobble is negligible in the present context.

In addition to its quasi-periodic wobble (annual + Chandler), the pole of rotation moves secularly at a rate of about 3.4×10^{-3} arc sec/yr, or 10 cm/yr (Dickman, 1977). This motion, if continued over a few million years, could amount to a polar displacement of large amplitude. Whether this occurs or not is not clear; most analyses of geomagnetic polar wander do not reveal much of a component that could be ascribed to large secular displacements of the rotational pole. But even if such large angular displacements did occur, it seems unlikely that they would require much energy. At constant angular momentum, it takes an amount of kinetic energy $\frac{1}{2}\omega^2 C(C - A)/A \cong 10^{27}$ J to displace the axis of rotation from the maximum to the minimum principal axis, for an undeformable earth. Since a major portion of the difference $C - A$ arises from the equatorial bulge that is itself an effect of rotation, it seems much more likely that in a deformable earth polar wander would be caused by incremental displacements of the principal axes of inertia, the pole of rotation remaining at all times very close to the axis of figure

C. Thus, polar wander, if it occurs, is unlikely to be an important energy sink.

SUMMARY

The earth is losing heat into space ("global heat flow") at a rate of 3×10^{13}–4×10^{13} W, the higher figure being now generally preferred. The heat flux across the earth's surface varies regionally by a factor of 2–3. Generally, heat flux is low in old continental shields and near oceanic trenches; it is high in continental regions of Cenozoic tectonic or volcanic activity and near oceanic ridges.

Heat is also brought to the earth's surface by lava, on land and on the oceanic floor. The total amount of the "volcanic" heat so carried is not precisely known, but is probably less than 8×10^{11} W. A certain amount of heat ("metamorphic heat") remains locked in those metamorphic rocks that form by endothermic reactions such as dehydration. It is estimated that in regions undergoing such metamorphism the heat flux rising into the crust from below may be at least twice normal; but since the area undergoing metamorphism at any one time is probably very small compared to the earth's surface area, metamorphic heat, like volcanic heat, is at most only a few percent of the global heat flow.

Tectonic phenomena, such as deformation and fracturing of rocks and uplift of mountains, require an input of mechanical energy. The rate at which potential energy must be supplied to uplift mountains is small ($\approx 7 \times 10^9$ W). Strain energy is converted to seismic energy at a rate probably less than about 1×10^{12} W. Since some strain energy is not released by faulting (e.g., the strain energy that goes into folding rather than into fracturing rocks), the rate at which strain energy accumulates is somewhat larger than 1×10^{12} W, but still probably constitutes no more than a few percent of the global heat.

Changes in the kinetic energy of rotation of the earth, from the secular tidal deceleration, are quite small. Tidal friction in the earth is a source of heat, not an energy sink.

Motion of plates requires no energy not already included in the global heat flux.

It would thus appear that we may not be grossly in error in assuming that the total power needed to drive all geological and geophysical processes is of the same order as the global heat flux, say 4×10^{13} W. But the energetic problem of the earth will not be solved by just pinpointing a heat source or sources capable of delivering 4×10^{13} W. This is where the concept of "structure," alluded to earlier, comes in. Suppose, for instance, that heat transfer in the earth is by conduction only, as for a long time it was assumed to be. Then, as is well known, the low thermal conductivity of rocks combined with the large dimensions of the earth makes it impossible for any large amount of heat to travel more than a few hundred kilometers in the earth's lifetime; thus a source of heat of the required intensity, but located in the core, could not account for the surface heat flow. On the other hand, for convection to occur certain conditions must be met, and in particular, the temperature gradient must exceed a certain critical value; this again places constraints on the location and distribution of heat sources. Finally, no heat source can account for the expenditure of strain energy, or for the motion of plates, without a suitable mechanism for converting heat into other forms of mechanical or potential energy. If that mechanism turns out to be convection, heat sources will have to be distributed so as to create the observed patterns of surface heat flow distribution and of plate motion.

We turn first to an examination of possible energy sources in the earth.

2 Energy Sources

Possible sources of energy in the earth may, for heuristic purposes, be classified as follows:

1. "Original" heat
2. Gravitational energy
3. Tidal dissipation of kinetic energy
4. Radioactivity

This classification is somewhat arbitrary, in the sense that, as we shall presently see, much of the original heat is probably gravitational energy released at the time of formation of the earth.

ORIGINAL HEAT

By original heat we mean the heat content of the earth very soon after formation. In the nineteenth century, cooling of an originally hot earth was thought to account for the whole of the surface heat flow as well as for the formation of mountains, believed to result from contraction of the cooling body. This hypothesis, which led, as is well known, to serious difficulties regarding the age of the earth, was gradually abandoned after the discovery of radioactivity, but some of it still survives, as there is no way of showing that the earth is not indeed cooling. As discussed in Chapter 4, secular cooling of the core is thought by some to provide, through the mechanism of crystallization of the inner core, the energy needed to generate the magnetic field.

The earth is now generally thought to have formed from the "solar nebula," a slowly rotating, disk-shaped mass of gas left over from the explosion of an earlier star in which heavy elements had recently been synthesized ("recently" here means not more than 10^7–10^8 yr before the formation of the earth). We picture the solar nebula, at the center of which the sun itself is in the process of forming, as cooling gradually by radiation, mostly in the direction normal to the plane of the disk; the radial temperature profile in the central plane of the disk is assumed to be adiabatic. The pressure also decreases radially outward and may be of the order of 10^{-3}–10^{-4} bar (1 bar = 10^5 Pa) at the radius of the earth's orbit. The composition of the nebula is assumed to be roughly the same as that of the sun, with hydrogen predominating.

The first stage in the formation of the earth and other planetary bodies consists in condensation of cooling nebular gas into solid "dust" particles, which then somehow come together in the second stage of accretion to form the planetary bodies. Condensation is reasonably easy to follow from thermodynamic data. The first solids to appear in the range 1750°–1600°K would be oxides, silicates, and titanates of calcium and aluminum (e.g., corundum, Al_2O_3; perovskite, $CaTiO_3$; melilite, $Ca_2Al_2Si_2O_7$). At a total pressure of 10^{-3} bar, metallic iron begins to condense at 1471°K. The bulk of the magnesian silicates (forsterite, enstatite) condenses around 1400°K. Oxidation of iron to the ferrous state (as in olivine, pyroxenes, etc.) and formation of FeS begin around 700°K (at $P = 10^{-4}$ bar), while hydrated silicates appear only below 500°K. From a variety of thermodynamic equilibria it has been inferred that the volatile-rich carbonaceous chondrites (a kind of meteorite) have formation temperatures in the range 300°–400°K. It would thus seem that since the earth obviously contains volatiles such as water and CO_2, the initial temperature of the material from which it formed could not have exceeded a few hundred degrees Kelvin.

This, however, is not necessarily the temperature of the earth itself just after forming, which depends on what happens during accretion. Accretion is much less easy to analyze than condensation. It is dominated by the consideration that a very large amount of gravitational energy must be released as particles fall onto the growing earth. This gravitational energy is of the order of 10^{32} J for the whole earth, quite enough to raise the temperature by tens of thousands of degrees and cause the earth to evaporate back into space as fast as it forms. The manner in which the earth manages to get rid of most of this energy, and in particular the fraction of it that is retained, determines the temperature and heat content of the growing planetary body.

Hanks and Anderson (1969) have made some calculations of initial temperature on the assumption that at every stage of accretion the surface temperature of the earth is such that it radiates energy back into the dust cloud at precisely the rate at which gravitational energy is released by dust particles free-falling onto its surface. There are several uncertain parameters in the relevant equation. The first is the temperature of the dust cloud itself, and the second is the rate of accumulation, or rate of growth of the earth's radius. This rate is unlikely to remain constant during the accretion process, since it would be essentially zero at the beginning, when the earth was so small that its gravitational attraction was very weak, and would be zero again at the end of the process, when accretion stopped because of depletion of the dust cloud. The mechanics of accretion are not sufficiently well known to allow a reliable estimate of the time required for accretion; estimates are commonly in the range 10^5–10^6 yr. There is also much debate as to whether accretion was heterogeneous or homogeneous. In the first case, it is supposed that accretion starts at essentially the same time as condensation. As the temperature drops in the nebula, the chemical composition of condensing particles changes, and so does the chemical composition of the growing planet. Thus iron is the first to aggregate, forming the earth's core before common silicates have begun to condense. The homogeneous model supposes, on the contrary, that condensation is complete before aggregation starts; the earth then has initially a uniform composition, and processes that occur later are required to separate the mantle from the core and the crust from the mantle. Hanks and Anderson (1969), assuming homogeneous accretion in 10^6 yr, calculate an initial temperature that barely exceeds 1000°K halfway down the mantle and drops off to about 500°K at the earth's center. Since present-day temperatures are far in excess of these values, it would seem that original heat cannot contribute much to the heat budget. Heterogeneous accretion, on the other hand, allows higher initial temperatures.

Considerable heating could also be generated, and higher internal temperatures reached, if the earth accumulated partly by the continuing capture of a planetesimal swarm of meteoritic bodies (Safronov, 1969; Wetherill, 1976, 1977). These bodies, hitting the earth at orbital velocities considerably greater than the velocity of free-fall, would cause much melting (as seems to have occurred on the moon) and would also, by means of shock waves, generate heat throughout the impacted body.

In brief, very little can be ascertained regarding the original heat content of, and temperature distribution in, the earth at the time of formation. Chemists assure us that the temperature of the original dust cloud must have been rather low, less than 1750°K, which is the temperature at which solid particles would begin to condense from the gaseous nebula. The subsequent events are quite uncertain and conclusions tentative. It seems nevertheless rather likely that initial temperatures were lower than they are today (see Chapter 3). Yet original heat is probably not entirely negligible. Sharpe and Peltier (1978) have recently worked out the thermal history of a convecting earth with a "crust" 64 km thick, in which radioactive sources presently generate one half of the surface heat flow, and with no deep-seated heat sources other than original heat. Although their model cannot be accepted literally, if only because there is surely radioactivity below 64 km, it nevertheless shows that original heat might indeed still contribute significantly to the surface heat flow.

GRAVITATIONAL ENERGY

A source of heat comparable in magnitude to radioactive heating (see below) is the release of gravitational energy inherent in changes in density distribution inside the earth.

The gravitational energy Ω of a body is defined as the sum of the gravitational interactions of all its past. Consider a shell of inner radius r, thickness dr, in a spherically symmetrical body in which the density ρ is a function of r only. This shell exerts no gravitational attraction on masses inside it, which in turn attract the shell as if their whole mass m_r were concentrated at the center, where $r = 0$; here

$$m_r = \int_0^r 4\pi \rho r^2 \, dr.$$

Thus the gravitational energy $d\Omega$ for the shell is

$$d\Omega = \frac{G m_r dm_r}{r},$$

where $dm_r = 4\pi r^2 \rho \, dr$ is the mass of the shell. Summing for all shells, we get

$$\Omega = G \int_0^R \frac{m_r dm_r}{r} = \frac{G}{2} \int_0^{M^2} \frac{d(m_r^2)}{r}, \qquad (2.1)$$

where R is the outer radius of the body and M is its mass.

Integrating by parts gives

$$\Omega = \frac{GM^2}{2R} + \frac{G}{2}\int_0^R \frac{m_r^2}{r^2}dr. \qquad (2.2)$$

Let ψ be the gravitational potential such that $g_r = d\psi/dr = -Gm_r/r^2$, where g_r is the gravitational acceleration at r. Then (2.2) can also be written as

$$\Omega = \frac{GM^2}{2R} - \frac{1}{2}\int m_r\, d\psi,$$

or

$$\Omega = \frac{1}{2}\int_0^M \psi\, dm_r.$$

The function ψ is, of course, a solution of Poisson's equation $\nabla^2\psi = -4\pi G\rho$.

If the body is in hydrostatic equilibrium, the pressure P at any point satisfies $dP = \rho\, d\psi$, and Ω can be written, from (2.1), as

$$\Omega = -4\pi\int_0^R r^3\rho\, d\psi = -4\pi\int r^3\, dP. \qquad (2.4)$$

If the pressure is zero on the outer boundary, integrating by parts gives

$$\Omega = 12\pi\int_0^R Pr^2\, dr = 3\int_0^V P\, dV, \qquad (2.5)$$

where V is the volume of the body. The apparent simplicity of (2.5) disguises the fact that P is a function of ρ and ψ. For a body with uniform density, $\Omega = 2GM^2/5R$. Equation (2.4) shows that Ω is easily calculated if the pressure distribution $P = P(r)$ is known. For the earth with the density distribution given by his model one, Birch (1965b) calculates $\Omega = 2.49 \times 10^{32}$ J.

CORE FORMATION

The change in density distribution most frequently considered corresponds to the separation of a dense core and lighter mantle in an earth formed by homogeneous accretion of material with uniform uncompressed density (uncompressed density means density at $P = 0$). The calculation is difficult, because any local change in ρ entails changes in g, ψ, and P everywhere; a change in P entails a further change in ρ because of compressibility. Furthermore, since the gravitational energy released is ultimately dissipated as heat, the temperature of the differentiated earth will be different from, and presumably higher than, that of the undifferentiated body; the effect of this change in temperature on the density must also be taken into account. Thus the equation of state $\rho = \rho(P, T)$ must be known. In a preliminary calculation, Birch (1965b) used for the undifferentiated state the pressure-density relation for the present (hot) earth and obtained for the gravitational energy $\Delta\Omega$ released by formation of the core $\Delta\Omega = 1.7 \times 10^{31}$ J. In a later paper (Flasar and Birch, 1973), a correction was made for the fact that the temperature would be lower in the undifferentiated earth than in the differentiated earth, precisely because of the release of gravitational energy; the corrected value of $\Delta\Omega$ now comes out at 1×10^{31} J, which is enough to raise the temperature of the whole earth by some 1500°C. Not all of this energy is available for heating, however; part of it (about 15 percent) goes into strain energy involved in compression under changing pressure. A negligible amount would go into kinetic energy of rotation, as the moment of inertia of the undifferentiated earth is larger than that of the differentiated one. Curiously, the radius of Birch and Flasar's undifferentiated earth is slightly smaller (6220 ± 20 km) than the present earth's radius, as if condensation of matter toward the center caused the earth to expand.

For comparison, note that 1×10^{31} J is more than the present total heat flux at the surface (4×10^{13} W) multiplied by the age of the earth. Formation of the core would thus be an energetic event of the first magnitude.

There is of course no unanimous agreement as to when this event took place. It might have occurred at the time of formation of the earth (heterogeneous accretion), which may have had a dense core from the start; in that case the $\Delta\Omega$ would be part of the 2.5×10^{32} J of the energy released at the time of formation, most of which was presumably lost by radiation from the accreting surface. More frequently, separation is assumed to have occurred at the time, of the order of half a billion years after formation, when radiogenic heating would have raised the temperature inside the undifferentiated earth to the melting point of iron, or of an Fe-FeS mixture, allowing liquid core material to trickle down much faster than solid material would. It has also been suggested that the process may be a slow, continuous one still in progress. Monin (1978), for instance, assumes that the rate of accretion at the core's surface is at all times proportional to the remaining average concentration of heavy material in the mantle; this leads to a core that is still growing and will continue to grow for another 1.9 billion years. Energy released to date is 1.61×10^{31} J, according to Monin, who uses a very approximate and unreliable equation of state. On the whole it appears unlikely that the core should still be growing, as its growth would be reflected in a secularly decreasing moment of inertia and a corresponding increase in the angular velocity of the earth, of which there is not much sign. Rates of rotation in the distant past, as determined from growth patterns of invertebrate shells, seem to be in good accord with deceleration rates from tidal friction, leaving but little room for any steady continuous acceleration. We also know, from paleomagnetic observations, that the earth already had a magnetic field of approximately the present intensity 2.6 billion years ago, a time when, according to Monin, the mass of the core would have been less than 15 percent of its final mass. Finally, seismic wave velocities and densities in the lower mantle do not seem to allow the presence in it of much metallic iron, except perhaps in the lower 100 km, just above the core boundary.

The density of the solid inner core being greater than that of the liquid outer core, progressive crystallization of a cooling core must release gravitational energy. It has been suggested (e.g., Braginsky, 1964; Gubbins, 1977; Loper, 1978) that this gravitational energy contributes significantly to the generation of the magnetic field. We shall return to this subject in Chapter 4.

TIDAL FRICTION

The gravitational attractions of the sun and moon produce a tidal deformation of the oceans and of the solid earth. Frictional dissipation of rotational kinetic energy accelerates the earth by an amount $\dot{\omega}_T$, a magnitude that has been a matter for debate for two centuries. The subject is by no means closed, as much uncertainty remains as to the magnitude of the tidal deceleration and the mechanism of dissipation, whether in ocean tides or in solid-earth tides. The acceleration $\dot{\omega}_T$ of lunar origin can, in principle, be determined from the acceleration $\dot{n}$ of the moon's orbital motion, if conservation of angular momentum for the earth-moon system is assumed. The lunar acceleration $\dot{n}$ can, again in principle, be determined from observations of the moon's position in ephemeris or atomic time. Until a few years ago, the accepted value of $\dot{n}$ was -22 arc sec/century², with a corresponding $\dot{\omega}_T = -5 \times 10^{-22}/\text{s}^2$. Modern observations have generally tended to increase (almost double) the value of $\dot{n}$ (Rochester, 1973), but a recent determination from lunar-range measurements puts it back to −24.6 arc sec/century² (Calame and Mulholland, 1978). These figures are difficult to check against the historical record of past eclipses because the earth's mantle also sustains accelerations from a variety of sources, including electromagnetic torques associated with fluctuations of the magnetic field. Yukutake (1972) has shown that the long-period ($\approx$8000 years) oscillation of the magnetic field may have caused in the last 2000 years a mantle acceleration of the order of $1 \times 10^{-22}/s^2$. The matter is further complicated by the possibility of a secular change in G, the gravitational constant.

Counting daily growth rings in fossil corals and other shell-building invertebrates has provided estimates of the number of days in a lunar month and in a year at known times in the geological past. When Newton reviewed the data in 1968 (Newton, 1968), he was led to conclude that G may be decreasing at a rate of the order of 1 part in 10^8 per century, while the moment of inertia C of the earth must have increased by a few percent in the last 350 million years. The uncertainties involved in counting growth rings are, however, so large that no great weight can be attached to these results.

It would seem now that all that can be said is that the tidal deceleration of the earth is probably between 5×10^{-22} and $10 \times 10^{-22}/s^2$, with a corresponding decrease in rotational kinetic energy in the range $3 \times 10^{12} - 6 \times 10^{12}$ W, roughly 10 percent of the present surface heat flow.

There has been a long-lasting debate as to whether dissipation, and hence heat production, occurs mainly in the oceanic tides (particularly shallow-sea tides) or in an imperfectly elastic mantle (solid-earth tides). Lambeck (1975) has reviewed the matter while calculating the acceleration of the moon's orbital motion from available models of the oceanic tides. The value of $\dot{n}$ he thus finds (-35 ± 4 arc sec/century²) is sufficiently close to recent astronomical determinations to allow him to conclude that a very major part, if not all, of the moon's acceleration is caused by dissipation in the oceans. The total loss of rotational energy is, according to him, about 5.7×10^{12} W, the oceans accounting for about 5×10^{12} W. The balance of 0.7×10^{12} W could be interpreted as a source of heat in the mantle but is hardly significant, as it represents the small difference between two large and uncertain numbers.

A well-known consequence of the tidal slowing of the earth's rotation is that the moon's mean distance c from the earth must increase at a rate that is presently of the order of a few centimeters per year. As the tidal dissipation varies as c^{-6}, it must have been much larger in earlier times. It is thus possible that in the early days of the earth-moon system, when the moon was much closer to the earth, internal heat generation by solid-earth tides may have been much larger than it is now. The early history of the earth-moon system is, however, not well understood. Although it is agreed that capture of the moon may have constituted a thermal event of the first magnitude for the earth, it is not agreed as to when capture occurred, if indeed it did. Just one more uncertainty in the thermal history of our planet!

RADIOGENIC HEAT

The important heat-producing radioactive isotopes in the earth are U^{235}, U^{238}, Th^{232}, and K^{40}. Other radioactive isotopes (e.g., Rb^{87}) decay so slowly, with such a small release of energy, and are present in such minute amounts that their contribution to the heat budget is negligible. The rate of heat generation is 0.97×10^{-7} W/g of natural uranium with the usual U^{235}/U^{238} ratio; 36×10^{-13} W/g of potassium, and $0.27\ 10^{-7}$ W/g of thorium. But how much uranium, thorium, and potassium is there in the earth?

The problem can be approached, but not solved, from different sides. One may assume, for instance, that the gross composition of the earth is the same as that of analyzed meteorites. But all meteorites do not have the same composition; eucrites, for instance, contain about 130 ppb uranium, while ordinary chondrites contain only 12 ppb. Alternatively, one could assess the uranium content of the upper mantle by measuring uranium in lavas that rise from the upper mantle (e.g., oceanic basalt), determining what fraction (5 percent?, 10 percent?) of the solid mantle melts to form these lavas, and more or less guessing by what amount uranium is enriched in the melt with respect to the solid from which it forms. Such calculations usually put the uranium content of the upper mantle at about 10–20 ppb. Measurements on lherzolite inclusions in some Australian lavas (Green *et al.,* 1968) give figures ranging, depending on the specimen, from 3 to 114 ppb, with an average for six specimens of 34 ppb. Larimer (1971) sets a lower limit for uranium in the earth as equal to the amount of uranium known or believed to be present in the crust, and an upper limit based on geochemical considerations regarding the degree of enrichment that is likely to occur in formation of the crust. He thus sets the uranium content of the earth at 12 ± 6 ppb. Ganapathy and Anders (1974) take the uranium content to be 18 ppb. Since the mass of the earth is close to 6×10^{27} g, its total uranium content would thus be about 1.1×10^{20} g, producing heat at a global rate of 1×10^{13} W. One would judge that the figure is uncertain by a factor of at least 2.

Once the uranium content has been selected, the thorium content follows from the observation that in most rocks and meteorites the Th/U atomic ratio is rather uniform, in the range from 3 to 4. Thus Ganapathy and Anders (1974) set the thorium content of the earth at 3.9×10^{20} g.

The problem of potassium is a difficult one. The abundance of a number of common elements (e.g., iron, calcium, aluminum, magnesium, and sodium) relative to silicon seems to be much the same on earth and in the sun and, by inference, in the original nebula from which the sun, earth, and other planets were formed. If the same rule applied to potassium, its abundance on earth should be about 800 ppm, which is close to the median for ordinary chondrites. Since, however, some elements, called "volatiles," are present in less abundance on the earth than in the sun, some fractionation must have occurred at the time of formation, and the question arises as to the degree to which potassium has also been fractionated. It has long been known that the ratio K/U is much lower on crustal rocks ($\cong 1 \times 10^4$) than in meteorites (8×10^4).

A minimum can be set for terrestrial potassium by noting that radioactive decay of K^{40} produces A^{40}, as well as Ca^{40}. On the assumption that all the A^{40} produced in the earth since its formation has now escaped into the atmosphere, knowledge of the amount of A^{40} presently in the atmosphere leads to an estimate of the minimum amount of potassium in the earth. This is a minimum because there is of course no guarantee that all the internally produced A^{40} has indeed reached the surface; it would in fact be surprising if it had, since volcanic gases today still contain nonradiogenic He^3 coming from the mantle (Craig *et al.*, 1978), where it must have resided ever since the formation of the earth. Various geochemical arguments regarding the probable degree of outgassing of the mantle thus lead to estimates of terrestrial potassium as 130 ppm (Larimer, 1978) or 170 ppm (Ganapathy and Anders, 1974).

But this is not the end of the story, because there is a possible connection between potassium and sulfur, first noticed by Lewis (1971). The S/Si ratio is considerably lower in crustal and upper-mantle rocks than in the sun, a fact that has been attributed to the chemical affinity of sulfur for iron and the consequent concentration of sulfur in the earth's core. It is indeed fairly certain that the density and sound speed in the core do not agree with the values expected for liquid iron under core conditions of pressure and temperature, and it has long been surmised that the core must contain substantial amounts (10–20 percent?) of an element much lighter than iron; sulfur, silicon, oxygen, and a few others have been suggested, with sulfur a favored candidate (see Chapter 4). Lewis, noticing that potassium tends to form strong bonds with sulfur (the compound K_2S, for instance, is quite stable), has suggested that the low potassium concentration of surface rocks could easily be accounted for without calling on complicated fractionation processes during condensation and accretion if some or most of the original potassium were now attached to sulfur and hidden from view in the earth's core, from which much of the radiogenic A^{40} would not yet have reached the earth's surface. Goettel (1972) has made some experiments to determine just how potassium would be partitioned between silicates and an iron sulfide melt. Oversby and Ringwood (1972) deny that such partitioning would occur to any appreciable degree, but Goettel (1976) insists that much of the terrestrial potassium could indeed be in the core, in which the potassium concentration would be of the order of 0.1 percent. His argument has received support from Bukowinski's (1976) calculations on the effect of pressure on potassium. Bukowinski shows that at core pressure, potassium would undergo an electronic transition that would effectively make it one of the transition elements, whose affinity for sulfur is well known. The matter of the concentration of potassium in the core is, as we shall see, of some interest in connection with the generation of the geomagnetic field.

Let us return now to radiogenic heat. The figures for potassium, uranium, and thorium proposed by Ganapathy and Anders (1974) are as shown in Table 2–1.

If we retain the same figures for uranium and thorium, but substitute 800 ppm for the abundance of potassium, the rate of heat generation comes to 3.8×10^{13} W, much closer to the 4×10^{13} W presently required.

The difference between the two models for potassium is very important. If the lower figure (170 ppm) is correct, there seems to be a good probability that the earth may be cooling, whereas Goettel's higher

TABLE 2–1 Abundance and Heat Generation of Potassium, Uranium, and Thorium

Element	Abundance (ppm)	Rate of heat generation for whole earth (W)
Potassium (K)	170	3.7×10^{12}
Uranium (U)	18×10^{-3}	1×10^{13}
Thorium (Th)	65×10^{-3}	1.05×10^{13}
Total		2.42×10^{13}

Source: Ganapathy and Anders (1974)

figure (800 ppm) allows for a steady state or, if gravitational heat sources are included (see below), for the possibility that the earth may be heating up. No definite statement can be made in that respect, the more so if we remember that the abundance of uranium and thorium in the earth may be uncertain by a factor of 2 or more.

CRUSTAL HEAT VERSUS MANTLE HEAT

It seems that observed local variations in heat flow at the surface of continents provide a datum by which crustal and mantle contributions to the heat flow can be separated. The observation is (Roy et al., 1972) that a plot of surface heat flow q against the rate of heat generation (radioactive) ϵ in the rocks in which the heat flow is measured shows a remarkable linear relation (Figure 2–1).

$$q = q_0 + b\epsilon \tag{2.6}$$

Here ϵ is measured on the same plutonic bodies in which q is measured; q_0 and b are two constants that characterize a region or geologic province. The heat flow q_0 is that which would be observed, in the same region, in near-surface rocks with $\epsilon = 0$; it represents therefore the heat coming from the lower crust and mantle. The constant b, which has the dimension of length, measures the vertical distance over which the heat sources are distributed. If q_0 were zero, and ϵ were uniformly distributed over a depth b, the steady-state heat flow would of course be precisely $b\epsilon$, as given by (2.6).

Lachenbruch (1968) was the first to point out that equation (2.6) seems to be obeyed regardless of the amount of erosion that has taken place since emplacement of the pluton in which q and ϵ are measured. This requires that ϵ be an exponentially decreasing function of depth z below the original surface. This follows, as noted by Albarede (1975), from the fact that if ϵ is a function of z, so is q; by differentiating (2.6),

$$\frac{dq}{dz} = b\frac{d\epsilon}{dz}. \tag{2.7}$$

But the continuity of heat flow requires, in the steady state, that div $\mathbf{q} = \epsilon$. With the one-dimensional geometry in which q depends on z only (and not on the other coordinates x and y) and with z counted positive downward, div $\mathbf{q} = -dq/dz$, so that, from (2.7),

$$\frac{dq}{dz} = b\frac{d\epsilon}{dz} = -\epsilon,$$

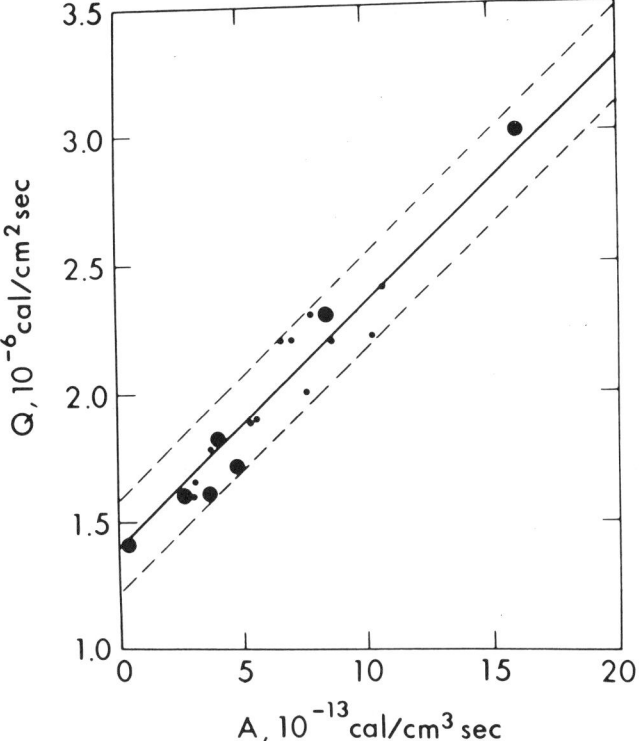

FIGURE 2-1   Heat flow Q plotted against radioactive heat production rate A for the southern Rocky Mountain region (large dots) and the Basin and Range province (small dots). Reproduced with permission from Roy *et al.* (1972).

which by integration yields

$$\epsilon = \epsilon_0 \exp(-z/b). \tag{2.8}$$

There is no generally recognized mechanism by which radioactive elements become heavily concentrated in the uppermost continental crust (see also Heier, 1978). What is of interest here is that the difference $q - q_0$ can be interpreted to represent the contribution of crustal radioactivity to the surface heat flow. This contribution turns out to be appreciable. Thus Roy *et al.* (1972) find that in the United States east of the Rockies the "reduced" heat flow q_0 from the deep crust and mantle amounts to no more than 0.8 HFU; corresponding figures for the Basin and Range province and for the Sierra Nevada are 1.4 and 0.4

HFU, respectively. By a slightly different method, Rybach *et al.* (1977) find that in the central Alps in Switzerland, only 0.6 HFU, slightly less than one half of the surface heat flow of about 1.5–1.7 HFU, enters the crust from the mantle where Moho is deepest ($\approx$55 km); mantle heat flow is slightly higher (0.8 HFU) under the foreland, where Moho is also somewhat shallower (40 km). In precambrian areas of Norway the average surface heat flow is 1.0 HFU and the reduced heat flow about 0.4 HFU (Heier and Grønlie, 1977). Thus one can probably estimate that radioactive heat sources in the crust generate about 50 percent of the continental surface heat flow. Since continents cover about one third of the earth's surface, roughly 15 percent of the global heat flow is thus accounted for.

SUMMARY

What emerges from this morass of fragmentary and uncertain data is that radioactivity by itself could plausibly account for at least 60 percent, if not 100 percent, of the earth's heat output. If one adds the greater rate of radiogenic heat production in the past by present radioactive elements to that of elements now extinct (Al^{26}, Pu^{244}), possible release of gravitational energy (original heat, separation of core, separation of inner core), tidal friction in the early days of the earth-moon system, and possible meteoritic impact in the early days of the earth itself, the total supply of energy may seem embarrassingly large. But, as we have repeatedly noted, most if not all of the figures mentioned above are uncertain by a factor of at least 2, so that disentangling contributions from the several sources is not an easy problem. It is, in fact, two problems, one in space (e.g., is there potassium in the core?), and one in time (when did the core separate from the mantle?). Not only must we know where the heat sources are, we must also know when they came into effect and what the rate of heat transfer may be. We must, in short, reconstruct the whole thermal history of the earth. Some clues may be provided by the present internal temperature distribution, a subject we turn to in the next chapter.

3 Temperatures Within the Earth

The present temperature distribution in the earth depends on (1) the original temperature distribution shortly after formation; (2) the distribution and intensity of heat sources, all of which are time dependent; and (3) the mechanism of internal heat transfer, whether by conduction or convection or both. Conversely, if the present temperature were known, it should be possible, at least in theory, to extract from it some information on, and possibly set upper or lower limits to, the distribution of internal heat sources.

Much effort has been devoted in the past to calculating the temperature distribution in the mantle, particularly the upper mantle, from the heat conduction equation. The main datum is the surface heat flow; a steady state is commonly assumed. Examples of such calculations are the often quoted "shield" and "oceanic" geotherms of Clark and Ringwood (1964) and the calculations by MacDonald (1965), which are now only of historical interest, if only because it is now certain that heat is also carried in the upper mantle by convection; not all heat transfer takes place by conduction in the radial direction. It is clear that, for instance, much of the heat flowing across the oceanic floor has been brought there by horizontal motion of hot material rising under, and moving away from, oceanic ridges. Curiously, it escaped everyone's (or almost everyone's) notice that these calculations were self-contradictory, insofar as the large horizontal temperature variations between subcontinental and suboceanic mantle that emerged from them (e.g., MacDonald, 1965) were sufficient to cause the very convection that had been assumed not to take place.

But when convection is included, there is no simple way of calculating the temperature, short of solving the full set of time-dependent equations expressing conservation of mass, momentum, and energy, which contain all the parameters of the conduction equation plus some poorly known parameters describing rheological properties and their dependence on pressure and temperature. Only partial solutions for idealized models, generally two-dimensional ones, are available at this time; there is still some debate as to whether convection embraces the whole mantle or is restricted to its upper few hundred kilometers. Furthermore, to get a solution one must first postulate what and where the heat sources are; thus we obviously cannot use these calculations to determine the heat sources, which is our main objective at this stage. We must thus turn to other, more direct methods for estimating what the temperature may be at a given place.

Broadly speaking, there are two methods for doing this. In the first method, one attempts to determine the temperature at a given depth from a known property of material at that depth. The property may be physical (e.g., velocity of elastic waves, electrical conductivity) or chemical (e.g., phase equilibrium between solid and liquid, or between polymorphs). In the second method, one tries to determine the temperature gradient in a homogeneous layer from an observed variation with depth of a physical property of the layer. In both methods, one must first allow for the effect of pressure, which, of course, also increases with depth.

Results, as we shall see, are on the whole rather disappointing. The main difficulty generally lies in estimating the effects of pressure. Pressure in the earth is tolerably well constrained, to better than 1 percent, but pressure coefficients of physical properties are hard to measure at the very high pressures of the earth's interior. Physical properties in the earth are generally not known very precisely. Density at a given depth, for instance, is not known to better than a few percent, which is also the magnitude of any likely temperature effect. The exact composition and nature of the mineral phases present in the lower mantle are still debated, and so is the composition of the core. Some physical properties (e.g., electrical conductivity) may be sensitive to variables (oxidation state, grain size, lattice defects, impurities) that, in the absence of representative samples, cannot be estimated at all.

We now review, in descending order from the top of the mantle to the inner core, some of the results.

THE UPPER MANTLE

THE LOW-VELOCITY ZONE

The low-velocity layer, as the name implies, is a layer in which the seismic velocities v_p and v_s are slightly lower than they are in the layers immediately above or below. The decrease in velocity with increasing depth is more pronounced for v_s (shear wave) than for v_p (compressional wave), and more pronounced under oceanic areas than under continents, where the low-velocity layer may either be absent or occur at greater depth. In oceanic areas the top of the low-velocity layer almost reaches the surface under ridges. Its depth increases progressively with distance from the ridge to a maximum of about 100 km; nowhere does it coincide with a sharp discontinuity. There is some evidence that the low-velocity layer coincides with the asthenosphere or "weak" zone, so named for its enhanced ability to flow as compared with the more rigid lithosphere above it. Lithospheric thickness is generally determined from seismic surface-wave dispersion; it appears to be well correlated with surface heat flow (Chapman and Pollack, 1977).

There are several possible interpretations for the low-velocity layer. Thomsen (1971) has shown that it could be due to a steep temperature gradient. The effects on seismic velocities of increasing temperature and increasing pressure being generally of opposite signs, the normal increase in velocity with increasing depth (pressure) could be reversed in a zone where the temperature gradient is sufficiently steep to overcome the pressure effect. The required temperature gradient is of the order of 5°–10°/km for v_s and 10°–20°/km for v_p.

At the moment, the preferred interpretation is that the reduced seismic velocities in the low-velocity layer are caused by the presence of a very small amount, of the order of 1 percent or so, of melt that forms a thin fluid film separating the grains of an otherwise solid rock. The evidence in favor of this interpretation is summarized by Solomon (1976). The temperature at the top of the low-velocity layer must then be the temperature at which melting begins at the corresponding pressure. The temperature of incipient melting depends of course on the composition of the rock but does not depart much from 1100°C (at $P = 1$ bar) for the several types of peridotite whose density and elastic properties make them likely constituents of the upper mantle. The melting point for dry peridotites rises with pressure at a rate of about 10°/kbar, bringing it to about 1400°–1500°C at a depth of 100 km, where the pressure $P \cong 30$ kbar. In the presence of water, however, the incipient melting point first decreases with increasing pressure, reaching a minimum of about 1000°C at 30 kbar for a water content of only 0.1 percent (Wyllie, 1971, Figure 6–18). The amount of water present in the mantle (presumably mainly in amphiboles and micas) is not exactly known. Carbon dioxide may also be present. Mysen and Boettcher (1975) have reported experiments on the melting of four different peridotites (two spinel and two garnet lherzolites) in the presence of water and CO_2 in variable proportions, and in the pressure range from 7.5 kbar to 30 kbar. When present alone, CO_2 does not lower much the solidus temperature from its value for the dry system, in contrast to water, which may lower the solidus to less than 900°C at 15–20 kbar. Mysen and Boettcher's experimental conditions are such that there is enough water present to saturate the melt at all pressures, a condition that may not be met in the mantle. Furthermore, the lherzolite specimens used in these experiments (inclusions in Hawaiian lavas and/or kimberlite pipes) may represent the fraction of the original mantle rock that is left after partial melting and removal of the most easily fusible components, and are therefore not necessarily typical of undepleted mantle or pyrolite.

What may be concluded from this is that if we accept the hypothesis that the top of the low-velocity zone or bottom of the lithosphere is indeed the depth of incipient melting, the temperature there is probably in the range 950°–1300°C where the lithosphere is 50 km thick, and in the range 900°–1400°C where its thickness is 100 km. In continental areas where the lithospheric thickness is 200 km, the temperature of incipient melting is probably not greater than 1200°C. The lower limit of the ranges quoted here applies when there is enough water present to saturate the melt, while the upper limit is for the "dry" case of no water at all.

If the bottom of the low-velocity zone is defined as the depth at which the shear velocity resumes the value it has at the top of the low-velocity zone, the thickness of the low-velocity zone comes out at about 100–150 km. The important point is that the thickness is finite, implying that at depths greater than 200–250 km the mantle is again entirely solid, or that the temperature gradient has dropped sufficiently for the effect of pressure on velocity to once more become preponderant, as it is above the low-velocity zone. Return to subsolidus temperatures could be caused either by a sharp drop in the gradient temperature, as in Figure 3–1a, by a sharp rise in the solidus temperatures, as in Figure 3–1b, or by a combination of the two. Melting in the presence of water entails a minimum solidus temperature.

Temperatures Within the Earth 33

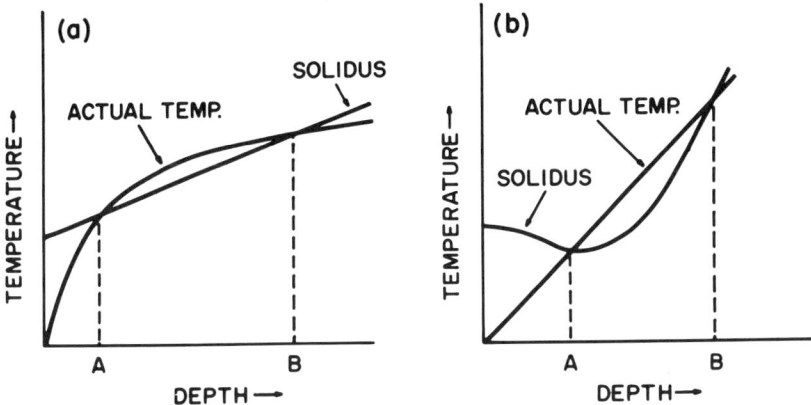

FIGURE 3–1 Possible temperature profiles in the upper mantle. The low-velocity layer extends from A to B. In (a), the temperature of incipient melting (solidus) rises linearly with increasing pressure or depth ("dry mantle"); the finite thickness of the low-velocity layer is due to a drastic reduction in slope of the temperature curve. In (b), temperature increases linearly while the solidus exhibits a minimum ("wet mantle"). The situation in the mantle is probably somewhere between cases (a) and (b).

GEOTHERMS FROM NODULES IN KIMBERLITE

Geotherms (i.e., temperature-depth curves) for subcontinental areas in the depth range 100–250 km have been estimated from the inferred temperature and pressure of equilibration of the several minerals present in lherzolite inclusions in kimberlites. The equilibrium distribution of the major elements calcium and magnesium between coexisting clinopyroxenes and orthopyroxenes in lherzolites is a function of temperature but is relatively insensitive to pressure. On the other hand, the distribution of aluminum between coexisting garnet and orthopyroxene is sensitive to pressure. The distribution coefficients are then compared with distribution coefficients measured on natural samples to determine the temperature and pressure at which the samples last equilibrated. Experiments are difficult because of the great length of time needed to reach equilibrium; the influence of other ubiquitous elements (e.g., iron) on distribution coefficients must also be ascertained.

Investigating in this manner a number of nodules from kimberlite pipes in Lesotho, Boyd and Nixon (1973) reported temperatures in the range 900°–1400°C at depths ranging from 140 km to a little over 200 km. The individual points fall nicely on two curves ("inflected" geotherms) indicating a sharp steepening of the temperature gradient at about 180 km. Specimens from below that depth are generally sheared, whereas those from above are granular. Boyd and Nixon point out that in the absence of marked differences in thermal properties above and below the inflection, continuity of steady-state heat flow across the inflection requires a continuous temperature gradient. They see in the inflected geotherm evidence for an event that caused major heating in the depth range 150–200 km. (The eruption of kimberlite that carried the nodules to the surface is in itself evidence for a perturbation of some sort, since kimberlite is not erupted continuously everywhere.) Figure 3–2 shows results of many investigators, as plotted by Harte (1978). Although there is much regional variation, the plot suggests possible temperatures of about 1000°C at 150 km, and of 1200°–1400°C at 200 km (65 kbar) and that the Mg/Fe ratio is 9; from Akimoto's data we would estimates for a given temperature, from 65 kbar to 40 kbar in one instance; corresponding depth estimates are reduced by 60–65 km. Harte (1978) suggests lowering the estimates shown in Figure 3–2 by perhaps 30–45 km. The matter can apparently not be resolved until we get more reliable experimental data on the distribution coefficients of major elements in coexisting phases. Perhaps all that can be said at the moment is that nodules in kimberlite suggest temperatures of about 1000° ± 200°C at 100-km depth, and 1200° ± 200°C at 150 km.

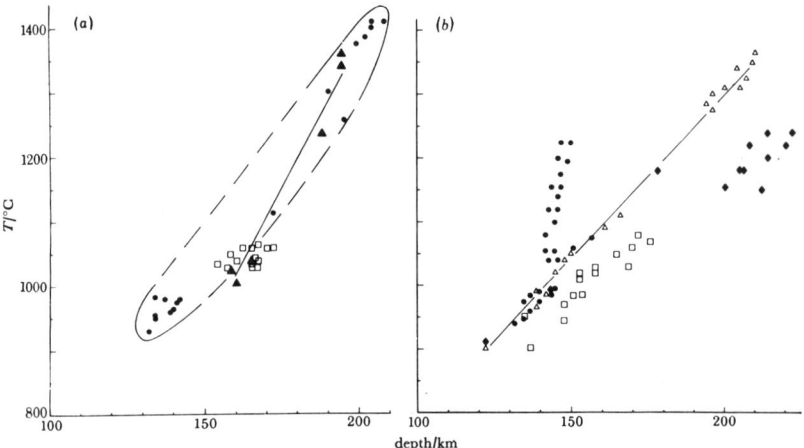

FIGURE 3–2 Estimated temperatures and depths of equilibration of garnet-lherzolite xenoliths in kimberlite pipes. Plot (a) is for pipes in northern Lesotho; (b) represents pipes in southern and southwest Africa and in Udachnaya (USSR). Symbols are as follows: triangles, Premier; squares, Kimberly; dots, Louwrencia; diamonds, Udachnaya. The depth scale may need reduction, in some instances by perhaps as much as 60 km (see text). Reproduced with permission from Harte (1978).

THE MANTLE'S TRANSITION ZONES

Ever since Bullen's early work (Bullen, 1936, 1940), seismologists have noted that the rate of increase with depth of the body-wave velocities v_p and v_s is particularly rapid in what Bullen called zone C, roughly between 350 and 900 km. Details of the seismic velocity distribution have long been, and to some degree still are, obscure. After much debate, it is now generally recognized that there is no discontinuity of the first order (discontinuous jump in velocity), or even of the second order (discontinuous change in velocity gradient), in this depth range. There do seem to exist, however, two narrow zones, at around 400 and 650 km, respectively, in which the velocity of both v_p and v_s increases with depth more rapidly than it does either immediately below or above these transition zones; the thickness and depth of these zones are still rather uncertain. Bullen had already recognized that velocities in zone C are incompatible with the assumption of a homogeneous mantle. Birch (1952) confirmed that heterogeneity might be due to phase changes rather than to changes in gross chemical composition.

Most of the minerals likely to be present in an upper mantle peridotite (olivine, pyroxenes, garnet) have now been found experimentally to undergo phase transformations at pressures such as exist in the mantle's C zone. For instance, at 1000°C olivine undergoes a major transformation at a pressure in the neighborhood of 110–120 kbar for olivines with a molar fraction of Mg_2SiO_4 (forsterite) greater than 0.8. The olivine transformation is now generally believed to account for the velocity transition zone near 400 km. Common olivine being a two-component system (Mg_2SiO_4-Fe_2SiO_4), any phase transition (e.g., melting) will take place, at fixed temperature, over a finite range of pressure. This spreading out of the transformation over a finite depth range accounts for the absence, near 400 km, of any first-order discontinuity in either density, elastic properties, or seismic velocities. The transformation has been studied experimentally by Ringwood and Major (1970) and by Akimoto (1972). Olivines with less than about 70 mol percent Mg_2SiO_4 invert to a so-called γ phase with the spinel structure; but magnesium-rich olivines invert to a phase (β) with a much distorted spinel structure that, according to Ringwood and Major, remains stable up to very high pressures. Akimoto, on the contrary, thinks that the β phase transforms to a γ phase at a pressure not greatly in excess of that of the olivine-β transition. At 1000°C, and for an olivine with 90 percent Mg_2SiO_4, the olivine-β transition begins at about 119 kbar, and the beginning of the β-γ transition is at about 135 kbar. With 80 percent Mg_2SiO_4, the olivine-γ transition begins at about

100 kbar; at 110 kbar the iron-rich γ phase inverts to a β phase that disappears again at 135 kbar. Pure Mg_2SiO_4 inverts to the β phase at 125 kbar and 1000°C; it is not known at what pressure it transforms to the γ phase, if indeed it ever does. From Akimoto's phase diagrams at 800° and 1000°C, rough values of the temperature coefficient of the transformation pressure can be determined. It appears to be about 45 bar/deg (dT/dP = 20.8°/kbar) for pure Mg_2SiO_4, somewhat less (30 bar/deg, dT/dP = 33.3°/kbar) for 80 percent Mg_2SiO_4.

The depths in the mantle at which the transformation begins and ends are not precisely known, and are likely to vary from place to place. Suppose that the transformation starts somewhere at 400 km (P = 132 kbar) and that the Mg/Fe ratio is 9; from Akimoto's data we would infer a temperature of about 1350°C. Ringwood and Major find a somewhat higher temperature (1600°C) at 400 km if that is the mean depth of the transformation zone and the olivine has 11 percent Fe_2SiO_4.

Considering the high uncertainty (50 percent) affecting the temperature coefficient of the transformation pressure, and our ignorance of the exact iron content of mantle olivines, of the precise depths at which the transformation begins and ends, and of whether the olivine-spinel phase transformation is indeed responsible for the seismic anomaly near 400-km depth, all we can say is that the temperature at that depth is probably in the range 1300°–1600°C. This is not a very precise estimate, but it is so far the best we have. It agrees with an earlier estimate by Graham (1970) of 1450° ± 120°C at 370 km.

Near 650-km depth ($P \cong 250$ kbar) there is another thin zone in the mantle in which the seismic velocities increase downward abnormally quickly. This has been variously attributed to a breakdown of olivine (spinel form) to $MgSiO_3$ + MgO (Liu, 1975b), to a transformation of pyroxenes to phases with ilmenite or perovskite structure, or to a breakdown of all silicates to their constituent oxides, with SiO_2 in the form of stishovite. The situation is summarized in Figure 3-3 taken from Liu (1977). None of the relevant phase equilibria is sufficiently well known to allow even an approximate determination of the temperature at which the phase change occurs.

THE LOWER MANTLE

LAYER D'

From differences in the gradient of seismic velocities v_p and v_s, Bullen (1950) divided the lower mantle into two zones, namely D' and D''. D' extends from 984 km to 2700 km and is characterized by a monotonic

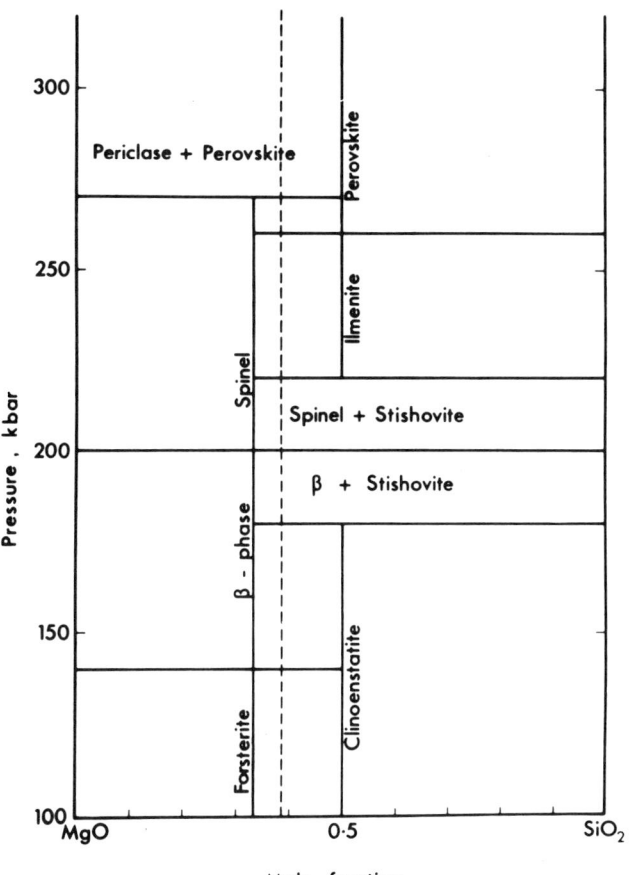

FIGURE 3-3 Schematic phase diagram for the system MgO-SiO₂ at about 1000°C. The dashed line represents a mixture 60 mol percent olivine plus 40 mol percent pyroxene. Reproduced with permission from Liu (1977).

downward increase in velocity. D'' extends from 2700 km to the core boundary (2885 km). It is now generally recognized that the velocity gradients are much smaller in D'' than in D' and may in fact be negative (Bolt, 1972; Cleary, 1974).

The lower mantle, from 700 km to the top of D'', is commonly considered to be mineralogically homogeneous even though some seismic observations have suggested to Johnson (1969) the possible presence of anomalous zones corresponding perhaps to phase transitions with small changes in density and elastic constants. For the time being, we shall ignore them and take D' to be homogeneous. We will return to the matter later.

In a homogeneous layer, in hydrostatic equilibrium, the variation in density ρ with depth is due to the effects of pressure P and temperature T. The pressure varies as $dP = -g\rho\, dr$, where r is distance from the center of the earth and $g = g(r)$ is the acceleration of gravity. The variation of density is then given by the relation

$$\frac{d\rho}{dr} = -\frac{g\rho^2}{K_T} - \alpha\rho\,\frac{dT}{dr}, \qquad (3.1)$$

where $K_T = \rho(\partial P/\partial\rho)_T$ is the isothermal incompressibility and $\alpha = -(1/\rho)(\partial\rho/\partial T)_P$ is the coefficient of thermal expansion. If α and K_T were known from the equation of state $\rho = \rho(P, T)$ for the material forming the layer, a knowledge of the density variation with depth $d\rho/dr$ would be sufficient to determine the temperature gradient dT/dr. The difficulty is that the equation of state for the lower mantle is not known *a priori*, since we aren't quite sure what the lower mantle is made of.

A major step was taken by Birch in his classical paper of 1952. An ingenious and rigorous transformation of Equation (3.1) leads to Birch's celebrated generalization of the Adams-Williamson equation:

$$1 - \frac{1}{g}\frac{d\phi}{dr} = \left(\frac{\partial K_T}{\partial P}\right)_T + T\alpha\gamma A + (T\alpha\gamma)^2 B + \frac{\alpha\phi\tau C}{g}, \qquad (3.2)$$

where ϕ is the seismic parameter $v_p^2 - (4/3)v_s^2$. The left-hand side of this equation is thus derivable from seismic observations. γ is Grueneisen's ratio

$$\gamma = \frac{\alpha K_T}{\rho c_V} = \frac{\alpha K_S}{\rho c_P} = \frac{\alpha\phi}{c_P}. \qquad (3.3)$$

K_S is the adiabatic (isentropic) incompressibility, and c_V and c_P are, respectively, the specific heat at constant volume and constant pressure; τ is the superadiabatic gradient defined by

$$\frac{dT}{dr} = -\frac{T\alpha g}{c_P} - \tau, \qquad (3.4)$$

where the first term on the right-hand side of Equation (3.4) is the "adiabatic" gradient derived from the thermodynamic relation for the change in temperature with pressure at constant entropy S,

$$\left(\frac{\partial T}{\partial P}\right)_S = \frac{\alpha T}{\rho c_P}. \tag{3.5}$$

In equation (3.2), A, B, and C are dimensionless functions of parameters such as $(\partial K_T/\partial P)_T$, $(\partial K_T/\partial T)_P$, $(\partial \alpha/\partial T)_P$, $(\partial c_P/\partial T)_P$; all these quantities are, in principle, derivable from the equation of state, provided that the equation of state be known in the pressure-temperature range of the lower mantle; then Equation (3.2) can, again in principle, be used to find T and τ. The difficulty is that the temperature terms, for instance $AT\gamma\alpha$, are small compared to the first term, $(\partial K_T/\partial P)_T$, which must therefore be known very accurately. As things stood in 1952, Birch concluded that all one could say was that a temperature of 5000°C produces no "serious discrepancy."

Since Birch's early work, some progress has been made in determining an equation of state applicable to the lower mantle. An interesting empirical discovery was made of a linear relationship between density and compressional wave velocity v_p, or bulk sound velocity $c = (K_S/\rho)^{1/2}$, for materials of the same mean molecular weight, regardless of crystal structure. Shock wave experimental data on a variety of minerals and rocks have also provided much information on density at very high pressures.

Graham and Dobrzykowski (1976), and more recently Watt and O'Connell (1978), have calculated the temperature distribution in the mantle between 700 and 1200 km that best fits density and seismic velocities, using third-order finite strain theory and assuming adiabaticity. Results depend on assumed composition (e.g., the ratio of (Mg,Fe)O to SiO_2), on mineralogy (e.g., mixed oxides versus phases with the perovskite structure), and, of course, on which of the many proposed density distributions is used. Graham and Dobrzykowski (1976) exclude pyroxene composition ((Mg,Fe)O/SiO_2 = 1), as it leads to "a temperature profile which is clearly outside the range of reasonable solutions" ($\approx$2500°C at 600 km). Peridotite and dunite compositions lead to a temperature of 1600° ± 400°C at the 671-km seismic discontinuity. Watt and O'Connell suggest temperatures about 1300°–1700°C at 700 km and an adiabatic gradient of about 0.3°/km.

In what is probably the best study of this kind to date, Wang (1972) used shock wave data to calculate temperatures in the lower mantle between 1300 and 2800 km. He starts from an acceptable density distri-

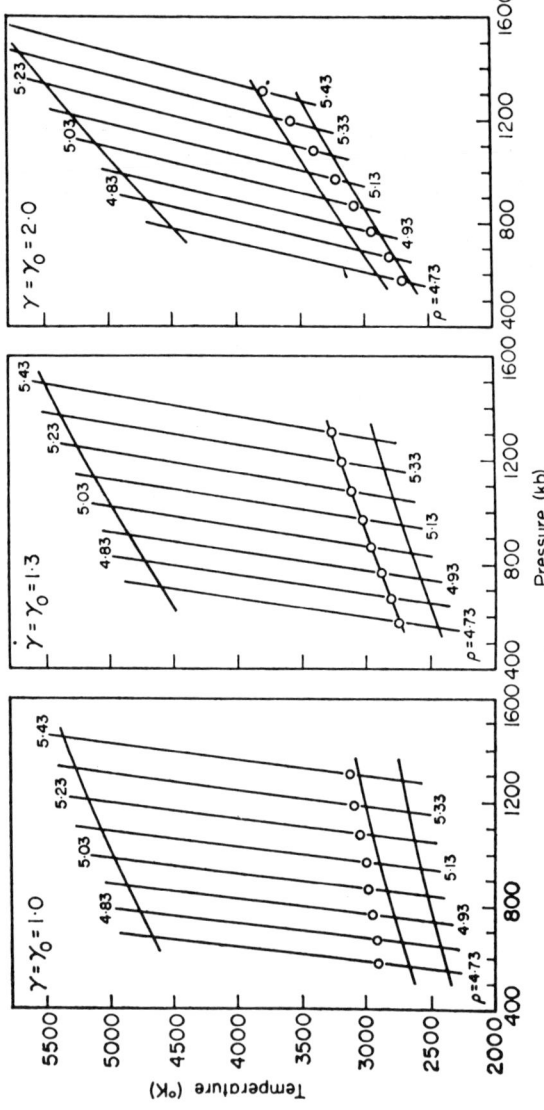

FIGURE 3-4 Isentropic curves (adiabats) for the lower mantle, for constant values of Grueneisen's ratio γ. The steeply inclined straight lines are lines of constant density; the circles are representative points for the lower mantle. Only for γ = 1.3 do these points fall on an adiabat. Reproduced with permission from Wang (1972).

bution that, when compared to the sound velocity c in the lower mantle, provides an estimate of gross chemical composition (mean molecular weight $m = 21.3$). By interpolation between shock-wave experiments, on dunite and fayalite, he determines a likely Hugoniot for material with $m = 21.3$; this gives him a density-pressure relation for mantle material at the temperature T_H of the Hugoniot.

Hugoniots can be reduced to isentropes if the Grueneisen ratio γ and its volume dependence (taken to be of the form $\gamma = \gamma_0 (V/V_0)^\eta$) are known. Wang chooses a set of likely values of γ_0 and η, and calculates for each set a number of isentropes, which he plots on a P-T diagram. Lines connecting points of equal density on different isentropes are drawn. Along any line of constant density ρ_i there is a point with pressure P_i corresponding to the depth in the mantle where ρ_i occurs; the temperature T_i at ρ_i and P_i is then read from the diagram (Figure 3-4). There is, of course, a different T_i for each set of values of γ_0 and η. Wang noticed, however, that only certain combinations of γ_0 and η (e.g., $\gamma_0 = 1.3$ for $\eta = 0$; $\gamma_0 = 1.5$ for $\eta = 1$) yield temperatures that, when plotted against P, fall on an isentrope. Thus the additional hypothesis that the lower mantle is isentropic permits selection of a preferred γ, and hence of temperature. The temperatures found in this manner do not strongly depend on the value of η, the difference between them being less than 100° at any depth. The mean of those curves shows a temperature of 2800°K at 1300 km, increasing almost linearly to 3300°K at 2800 km with an average gradient of about 0.33°/km. Wang estimates the uncertainty on the temperature to be less than ±800°.

Wang's results extrapolated upward give a temperature of about 2000°C at 700 km, which is also the upper limit of Graham and Dobrzykowski's (1976) estimate of 1600° ± 400°C.

All the previously mentioned studies assume an isentropic (=adiabatic) lower mantle. The reason for doing so is as follows. The viscosity of the lower mantle has now been shown to be of the order of 1×10^{22} poise (Cathles, 1975; Peltier, 1976). As discussed in Chapter 5, this relatively low viscosity corresponds to a Rayleigh number much in excess of that needed for convective instability. Transport of heat by convection is so much more effective than transport by conduction that convection, when it starts, will tend to reduce the vertical temperature gradient to the minimum value needed to maintain convection, which is precisely the adiabatic gradient. Numerical solutions to the equations governing steady-state convection (e.g., Turcotte and Oxburgh, 1967) indeed commonly show a convective pattern consisting of a thick adiabatic core between thin top and bottom boundary layers, so that the average temperature gradient over a large fraction of the volume of the fluid does not depart significantly from its adiabatic value. But should we discover that the gradient in the lower mantle is significantly greater or smaller than the adiabatic, we would have to conclude that the lower mantle is not convecting; this would lead to serious difficulties regarding transfer through the mantle of the large amount of heat that, as we shall show in Chapter 4, must be leaving the core.

Somerville (1977) has recently re-examined in much detail the properties of the lower mantle and has extracted values for the superadiabatic gradient τ from Birch's (1952) generalization of the Adams-Williamson equation,

$$\frac{d\rho}{dz} = \frac{\rho g}{\phi}\left(1 - \frac{\tau\alpha\phi}{g}\right), \qquad (3.6)$$

where z is depth and the other symbols are as before. He selects a number of density models satisfying the usual constraints for total mass, moment of inertia, and normal oscillation modes, each model leading to a different τ. (The different models selected from the literature generally differ from each other, as regards density at a given depth, by no more than 1 or 2 percent.) The parameter ϕ is known from seismic data, and ρ and $d\rho/dz$ are taken from the model; g is also calculated from the density model. Thus τ could be calculated directly from (3.6) if α were known. There is no direct way of determining α for the lower mantle; usually one starts from (3.3),

$$\alpha = \frac{\gamma c_P}{\phi},$$

noting that at high temperature c_P is insensitive to either P or T, and that quantities related to, but not exactly equal to, Grueneisen's ratio γ can be derived from seismic data (Verhoogen, 1951) by use of Debye's theory of specific heat or the Mie-Grueneisen equation of state.

Noting that $c_P = c_V (1 + T\alpha\gamma)$, and hence $\alpha = \gamma c_V/(\phi - T\gamma^2 c_V)$, Somerville calculates α by taking $\gamma = \gamma_D$, γ_D being the "Debye frequency" or "acoustic" Grueneisen parameter,

$$\gamma_D = -\frac{V}{\theta}\frac{d\theta}{dV}, \qquad (3.7)$$

where V, the specific volume, is $1/\rho$; θ is the Debye temperature. The Debye temperature in the deep mantle is calculated from Debye's formula

$$\theta = \frac{h}{k}v_D \tag{3.8}$$

(h = Plank's constant, k = Boltzmann's constant), where v_D, the cutoff Debye frequency of the vibrational spectrum, is

$$v_D = \left(\frac{9N}{4\pi V}\right)^{1/3} \bar{v}, \tag{3.9}$$

with $1/\bar{v}^3 = (1/v_P^3) + (2/v_s^3)$. Here N/V is the number of atoms for unit volume; while v_P and v_s are, respectively, the velocities of the compressional and shear elastic waves. The heat capacity at constant volume c_V has its classical Debye value slightly corrected, by empirical means, for anharmonicity of lattice vibrations. As a starting point, a temperature of 1900°K at a depth of 670 km is assumed. This temperature was first suggested by Ahrens (1973). Starting from an upper mantle consisting of garnet, orthopyroxene, clinopyroxene, and olivine, Ahrens first computed the composition of the phases that would be in equilibrium at various depths, using Akimoto's (1972) phase diagram for the olivine-spinel transformation. Likely values of the seismic velocity v_p were then computed from laboratory ultrasonic experiments. The iron content, proportion of garnet, and temperatures were then adjusted to give the best fit to the actual v_p distribution down to 670 km.

Somerville finds that (1) the adiabatic gradient in the lower mantle is typically 0.4°/km, (2) the superadiabatic gradient varies from +0.5 to −0.8°/km, depending on the density model used at the start. It is very close to zero for Jordan and Anderson's (1974) model B. If $\tau = 0$, the temperature at 2770 km then comes out at 1900° ± 0.4 × 2100° = 2740°K, or roughly 2500°C.

Somerville has also examined multilayer models of the lower mantle in which part of the density increase with depth is attributed to phase transformations rather than to elastic compression. The superadiabatic gradients for multilayer models range from 0.6°/km to 1.4°/km. Note, however, that since the nature of the phase transformations is not known, it is impossible to estimate the corresponding entropy change; a "superadiabatic" gradient in Somerville's sense does not necessarily imply that the lower mantle is not isentropic. All that can be said is that the temperature gradient in the lowermost mantle is likely to be higher by about 1°/km if the mantle is multilayered; this would add about 2000° to the temperature at 2800 km. We note, however, that the layered structure of the lower mantle is not universally accepted by seismologists.

It is difficult to assess the confidence to be placed on these results. Much depends, in Somerville's work, on the identification of Grueneisen's ratio γ with γ_D, the acoustic Grueneisen parameter derived from Debye's theory. Debye's theory applies only to a "harmonic" solid, that is, a solid in which a displacement of an atom from its equilibrium position is resisted by a force strictly proportional to the displacement; such a Debye solid has $\alpha = \gamma = (\partial K_T/\partial P)_T = 0$. To account for finite thermal expansion, for the dependence of γ on volume, and for the observed temperature dependence of K_T, it is necessary to use "fourth order" theory, that is, to retain all terms up to order 4 in the Taylor expansion of the interatomic potential function (Debye's theory is a second-order theory) (Thomsen, 1971). To obtain a valid fourth-order equation of state from which α could be calculated at any T and P, it is necessary to measure precisely at least six quantities (e.g., ρ, α, K, $(\partial K/\partial P)_T$, $(\partial^2 K/\partial P^2)_T$, $(\partial K/\partial T)_P$), some of which (e.g., the second derivative of the incompressibility) are not readily determined; furthermore, the measurements must be made on the same high-pressure phases that are present in the mantle, the nature of which is still largely unknown. Yet it would seem, in spite of all the uncertainties, that the adiabatic gradient in the lower mantle cannot be far from 0.3°–0.4°/km on the average. Wang's estimate of 3300° ± 800°K at a depth of 2700–2800 km seems realistic.

LAYER D''

We now consider layer D'', comprising the lowest 100–200 km of the mantle. The lower boundary of D'' is the core-mantle boundary at a depth of approximately 2900 km; the upper boundary of D'' is not sharply defined. Recall that layer D'' is characterized by a small and possibly negative seismic velocity gradient for both v_p and v_s. Bolt (1972) interpreted the seismic data as requiring an anomalous downward density increase that he attributed to admixture in increasing proportion of core material. Jones (1977) has shown that the seismic data could also be accounted for by a temperature gradient in D'' of about 12°/km.

This high gradient, which contrasts with the much smaller adiabatic gradient of 0.3°–0.4°/km prevailing in the lower mantle just above D'', may be an effect of the relatively large amount of heat that seeps from the core into the mantle. As will be shown later, generation of the earth's magnetic field almost certainly requires that an amount of heat variously estimated as between 2×10^{12} and 10×10^{12} W (see Chapter 4) be carried, mostly by convection, to the core's surface and into the mantle. If a commonly accepted (1.4×10^{-2} cal/cm s deg, or about 6 W/m deg) value of the thermal conductivity of the lower mantle is used, it is easy to calculate that to carry, say, 8×10^{12} W through the mantle by conduction alone, a temperature of about 14,000°C is required at the core-mantle interface. Since this is much more than the melting point of mantle material (corrected for pressure) and is inconsistent with the existence of a solid inner core, which requires the temperature at the boundary of the inner core to be less than some 8000°C (see below), it follows that transfer of heat through the lower mantle must be by convection. But the core and mantle are so different with respect to properties that govern convection (e.g., Prandtl number) that their flow patterns must necessarily also be very different, requiring the presence between them of a temperature boundary layer in which steep and horizontally variable temperature gradients allow for continuity of temperature and heat flow. The D'' layer is thought to represent this thermal boundary layer (Verhoogen, 1973; Jones, 1977; Elsasser et al., 1979). If the interpretation is correct, the temperature T_c at the core-mantle boundary would exceed the temperature at depth 2800 km by some 1200° if D'' is assumed to be 100 km thick and if Wang's estimate is accepted.

Much of the interpretation of D'' as a thermal boundary layer with a steep temperature gradient rests, however, on the assumed value of the thermal conductivity of the lower mantle, k, which is, unfortunately, one of the less well known geophysical parameters.

The value of k quoted above (1.4×10^{-2} cal/cm s deg, or 6 W/m deg) is taken from Horai and Simmons (1970), who calculated it in the following manner. They first note an empirical relation between k and seismic velocities for silicate minerals:

$$v_p = (0.17 \pm 0.02)k + (5.93 \pm 0.17),$$
$$v_s = (0.09 \pm 0.02)k + (3.31 \pm 0.16), \tag{3.10}$$

where v_p and v_s are expressed in km/s and k is in mcal/cm s deg. (Applied to the lower mantle at 2700 km, where $v_p \cong 13.6$ and $v_s \cong 7.26$, these relations give, respectively, $k = 45.1$ and $k = 43.9$; but see below.) Noting the relations (3.8) and (3.9) between seismic velocities and Debye temperature θ, Horai and Simmons proceed to transform equations (3.10) into an empirical relation between k and θ:

$$\theta = (25.6 \pm 3.0)k + (385 \pm 28) \tag{3.11}$$

The Debye temperature for the mantle is calculated from seismic velocities, using (3.8) and (3.9); this gives, at 2898 km, $\theta = 1242°K$, and $k = 34$. They then proceed to correct for the temperature (unspecified by Horai and Simmons, but presumably room temperature) at which the empirical relations (3.10) and (3.11) are found to be valid. The temperature correction is based on theoretical work that predicts that k will be proportional to

$$\frac{\theta^3}{\gamma^2 \rho^{1/3} T} \quad \text{(for } T > \theta\text{)},$$

where γ is Grueneisen's ratio and ρ is the density. The corrected value of k at 2898 km comes out, according to Horai and Simmons, at 14 mcal/cm s deg; this is about one third of the value calculated by substituting directly in (3.10) the observed seismic velocities of the lower mantle.

Horai and Simmons' result may be open to criticism. For one thing, it is difficult to know what significance should be attached to the empirical relations (3.10). Thermal conductivity is indeed proportional to the velocity of phonons, which is an averaged sound velocity, but k depends also on factors unrelated to elastic-wave velocities. At $T > \theta$, the thermal conductivity of a perfect crystal of large dimensions depends mostly on phonon-phonon scattering, which is a function of the anharmonicity of lattice vibrations; a crystal in which lattice vibrations are purely harmonic has infinite thermal conductivity as well as zero thermal expansion, and $\gamma = 0$. Most theories of lattice conduction (see, for instance, Drabble and Goldsmid, 1961) take anharmonicity into account by introducing Grueneisen's ratio into the formulation; this leads, for instance, to Lawson's formula (Lawson, 1957)

$$k = \frac{r_0 K_T^{3/2}}{3\gamma^2 \rho^{1/2} T}, \qquad (3.12)$$

where r_0 is the nearest-neighbor distance and K_T is the isothermal bulk-modulus. Mao (1973a) has used (3.12) to predict that a lower mantle consisting of dense oxide phases (e.g., stishovite, magnesiowustite) with high incompressibility would have a thermal conductivity approaching 1 cal/cm s deg (418 W/m deg), or 70 times Horai and Simmons' corrected value! Kieffer, using yet a different method though assuming the same composition (periclase + stishovite), estimates the lattice conductivity at the core-mantle boundary to be about

10 mcal/cm s deg (Kieffer, 1976), in broad agreement with Horai and Simmons' value.

Add to this uncertainty the possible (but unknown) effects of phonon scattering by crystal defects and grain boundaries, and add a (possibly small) contribution to the conductivity from radiative transport of heat (Mao and Bell, 1972; Mao, 1973b; Schatz and Simmons, 1972) or from electronic excited states (Mao, 1973a), and it becomes clear that k is not known to better than one order of magnitude, with Horai and Simmons' value erring possibly on the low side. Note that if we accept Jones' (1977) interpretation of the seismic anomaly in layer D'' as being caused by a steep temperature gradient of 12°/km, and assume the heat flow Q from the core to be 8×10^{12} W, the thermal conductivity required to maintain that gradient ($k = -(Q/4\pi r_0^2)\, \partial T/\partial r$, where r_0 is the core radius) turns out to be 4.4 W/m deg (10.5 mcal/cm s deg), close to Horai and Simmons' estimate of 14 mcal/cm s deg. Mao's (1973a) value $k = 1$ cal/cm s deg would require, to maintain a gradient of 12°/km, a heat flux into the mantle Q of 1.82×10^{14} cal/s = 7.6×10^{14} W, many times the earth's total heat output and therefore very unlikely; if Mao is right, the seismic anomaly in layer D'' is not an effect of temperature, and the temperature is essentially the same at 2700 km and at the core boundary.

If k in the lower mantle is about 10 mcal/cm s deg, the conducted heat flow corresponding to the gradient above D'' (0.4°/km) is 4×10^{-8} cal/cm s. Thus either the heat inflow from the core is 30 times less than 2×10^{12} cal/s (8×10^{12} W), which corresponds to a heat flow at the core boundary of 1.25×10^{-6} cal/cm² s, or the mantle convects with a Nusselt number N of the order of 30. (The Nusselt number is the ratio of total heat carried to heat carried by conduction alone.) The latter hypothesis is acceptable. Turcotte and Oxburgh (1967) have shown that the convective pattern in a viscous layer of thickness d heated from below will consist of a thick isothermal core between two horizontal boundary layers of thickness δ, with rising and falling plumes also of thickness δ. If R is the Rayleigh number, they predict

$$\delta \cong dR^{-1/3}, \qquad w \cong \frac{\kappa R^{2/3}}{d}, \qquad N \cong R^{1/3},$$

where w is the vertical velocity and κ is the thermal diffusivity, $\kappa = k/\rho c_P$. If k is about 10^{-2} cal/cm s deg, κ is $\approx 6 \times 10^{-3}$ cm²/s. We want a boundary layer thickness $\delta \cong 100$ km, which for $d = 2900$ km requires $R^{1/3} \cong 29$, and consequently $N \cong 29$; the corresponding vertical velocity w is 0.55 cm/yr. Similarly, Moore and Weiss (1973) find, for convection dominated by viscosity (high Prandtl number),

$$N \cong 2(R/R_c)^{1/3},$$

where R_c is the critical Rayleigh number at which convection starts. Since R_c for the mantle is $\approx 10^3$–10^4, a Nusselt number of 30 requires a Rayleigh number of order 3×10^6–3×10^7, which does not appear impossible. (See also McKenzie et al., 1974.)

To recapitulate, the interpretation of the seismic anomaly in D'' as an effect of temperature requires a temperature gradient in D'' of about 12°/km (Jones, 1977). This gradient equals $q_0/k = Q/4\pi r_0^2 k$, where Q is the total heat escaping from the core. If $k = 10^{-2}$ cal/cm s deg (4.18 W/m deg), $Q = 2 \times 10^{12}$ cal/s $= 8 \times 10^{12}$ W. There seem to be no major inconsistencies in these figures, or between them and the existence of an adiabatic gradient of 0.3–0.4°/km in most of the lower mantle; in particular, the Nusselt number of about 30 predicted from these figures for the lower mantle seems in line with various theoretical estimates. The temperature at the core-mantle boundary is then 4500°K if we start from Wang's estimate (Wang, 1972), or close to 6000°K if we take Somerville's result for a multilayered lower mantle.

Much depends, however, on the value of k. It is not impossible, for theoretical reasons (Mao, 1973a), that k could be much larger than the value adopted here. In that case, interpretation of D'' as a thermal boundary layer would have to be abandoned, since it would require Q to be equal to or greater than the total heat flux at the earth's surface. Supposing that the temperature gradient in D'' is the same as in the rest of the lower mantle, the temperature at the core-mantle boundary is then approximately 3350°K (Wang), or 4800°K (Somerville, for layered mantle).

Perhaps some support of the interpretation of D'' as a thermal boundary layer with a steep temperature gradient may be provided by the observation that the temperature at the core-mantle boundary is not likely to be as low as 3350°K, judging from the fact that the outer core is liquid. As will be discussed in greater detail in the next chapter, the melting point of iron at the pressure (1.4 Mbar) of the core-mantle boundary cannot be much less than 5000°K (Leppaluoto, 1972a,b). Addition of sulfur lowers the melting point. Very little can be said on the matter, since the exact sulfur content of the core (10 percent, 15 percent?) is unknown, as is the phase diagram for the Fe-FeS system at the relevant pressure. A sulfur content of 15 percent lowers the liquidus by about 200° at ordinary pressure and probably by not more than 1000° at $P = 1.4$ Mbar; thus the liquidity of the core seems to require a temperature not lower than 4000°K at the core-mantle boundary.

THE OUTER CORE

When it comes to guessing at the temperatures in the core, roughly three lines of approach are used:

1. Determine the temperature from seismic data, an acceptable density distribution, and an equation of state.
2. Assume the temperature to be known at the core-mantle boundary (CMB), and assume further the outer core to be isentropic. Temperatures then fall on the adiabat through the known point. This calculation requires only Grueneisen's ratio γ.
3. Assume the temperature at the inner core–outer core boundary to be the melting temperature of inner-core material at the relevant pressure. Since the core does not consist of a pure substance (e.g., iron), melting temperatures depend on composition, a topic to which we now turn.

COMPOSITION OF THE CORE

It has been known for some time that the density of the outer core is noticeably less than the density predicted from shock wave experiments on iron. Neither do experimental sound velocities at a given density agree with observed velocities in the core (Birch, 1963).

Elements besides iron that could be present in the core in sufficient proportion to reduce its density by about 10 percent are few: oxygen, magnesium, silicon, sulfur (carbon and hydrogen are readily soluble in iron but do not much affect its density). Silicon, once considered a likely candidate, now seems to have fallen out of favor, mainly because of its high affinity for oxygen. It is indeed difficult to see how the highly reducing environment required by the presence in the core of metallic silicon could be compatible with the oxidizing conditions implied by the presence of FeO in the mantle. When in contact, silicon in the core would reduce FeO in the mantle to form SiO_2 (or a silicate) and metallic iron.

Alder (1966) made some rough thermochemical calculations suggesting that MgO could be soluble in liquid iron under the pressures and temperatures of the core. This suggestion has not been followed up, perhaps for lack of experimental data. Oxygen (as FeO) may be present in the core in sufficient amount to account for its density (Dubrovskii

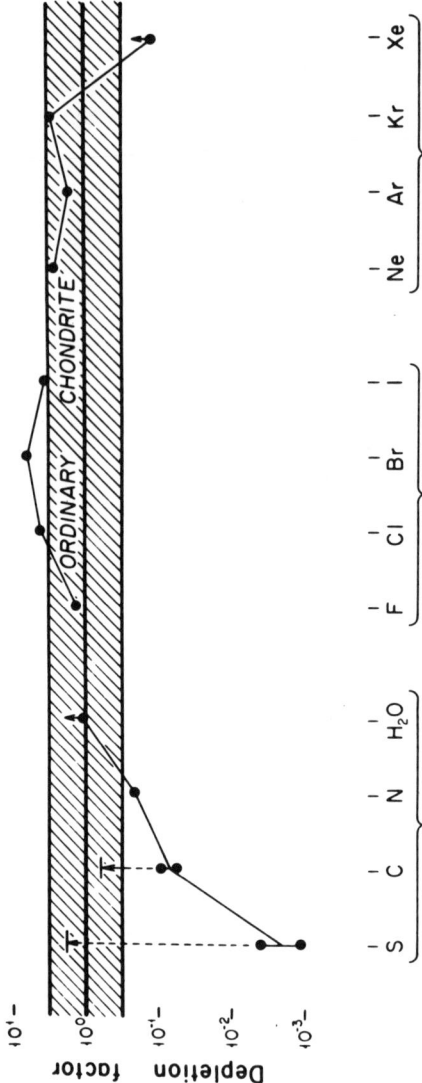

FIGURE 3-5 Abundances of some volatile elements in crust plus mantle normalized to their abundance in chondrites. The two values given for sulfur and carbon are for different estimates of crustal abundances of these elements. The severe depletion of sulfur in the earth's crust and mantle as compared to chondrites is an argument in favor of the presence of sulfur in the earth's core. Reproduced with permission from Murthy and Hall (1970).

and Pankov, 1972). Ringwood, once a staunch supporter of silicon has recently endorsed this view (Ringwood, 1977), which again lacks experimental support. At ordinary pressure, a mixture of metallic iron and iron oxides containing, say, 10 percent oxygen would melt at a temperature only a few degrees below the melting point of pure iron to produce two immiscible liquids, one of which is a metallic solution containing very little oxygen (approximately 0.2 percent), while the other has a composition close to FeO. Although the solubility of oxygen in the metallic liquid increases rapidly with increasing temperature, it is not known at what temperature the immiscibility gap vanishes, or even whether it closes at all. Pressure would presumably increase the solubility of FeO. Ringwood estimates that above 500 kbar, complete miscibility would occur. The problem is further complicated by the tendency of Fe^{2+}, the usual form of iron in FeO, to disproportionate at high pressure (Mao and Bell, 1977):

$$3\,Fe^{2+} = 2\,Fe^{3+} + Fe\,(metal).$$

The arguments for or against the presence of sulfur in the core are mainly based on consideration of natural abundances, apart from the well-known affinity of iron for sulfur. In the crust and mantle, there is very little sulfur (about 600 atoms per 10^6 atoms of silicon), in sharp contrast to chondritic meteorites (about 10^5 atoms for 10^6 of silicon) or to the sun, where the atomic ratio of sulfur to silicon is somewhere between 0.2 and 0.6, with a probable value of about 0.35 (Ross and Aller, 1976). For the earth to have the same S/Si ratio as the sun, the earth's core would have to contain about 18 percent sulfur by weight. While there is no reason to expect precisely the same ratio in the sun, earth, and meteorites, it is difficult to explain, if the core contains no sulfur, why the earth should be so much more strongly depleted in sulfur than in the other "volatile" elements (carbon, nitrogen, fluorine, chlorine, bromine, and iodine and the rare gases neon, argon, krypton, and xenon), relative to chondritic meteorites (Murthy and Hall, 1970, 1972), which are themselves not significantly depleted in sulfur with respect to the sun (Figure 3–5). Many decades ago, Goldschmidt had stated on purely geochemical grounds that the earth must contain much unseen sulfur in the form of FeS, a suggestion that was later revived by Mason (1966) and Murthy and Hall (1970). Arguments based on abundances are, however, not entirely convincing because, as Brett (1976) noted, the sulfur content of the mantle is essentially unknown except for its upper 100 or 200 km, which might not be representative of the whole. More convincing arguments are those of Murthy and Hall (1970,

1972) and Murthy (1976), who argue that in any model of homogeneous accretion of the earth, sulfur is bound to follow iron into the core, if only because the Fe-FeS eutectic liquid is the first liquid to form, at the lowest temperature, for all reasonable terrestrial compositions. If there is any sulfur at all in the earth, some of it is bound to be in the core.

It is, of course, possible, and even likely, that the earth's core might contain, in addition to sulfur and to nickel, which is ubiquitous in iron meteorites, some proportion of carbon, magnesium, oxygen, and still other elements. As we shall see later, there is a fair chance that it might contain small amounts, of the order of 0.1 percent or less, of potassium. But as something is known of the phase relationships in the Fe-FeS system, while nothing (or almost nothing) is known of the Fe-FeO or Fe-MgO systems at very high pressure and temperature, we will assume in what follows that the earth's core consists essentially of Fe + FeS, with an FeS/Fe ratio somewhere near 10–15 percent. We now return to the matter of determination of temperature by the three methods outlined above.

TEMPERATURE FROM AN EQUATION OF STATE

The problem has been examined with much care by Stewart (1973), who proceeds as follows: Assume that the density, the pressure, and the seismic parameter ϕ (which is the square of the sound velocity, $\phi = K_s/\rho$) are known at all depths in the outer core (but note that the density is usually calculated, as by the Adams-Williamson equation, on the assumption of adiabaticity). Assume that the Grueneisen ratio varies as

$$\gamma = \gamma_0 \left(\frac{V}{V_0}\right)^\eta,$$

where the subscript 0 refers to zero pressure and V is specific volume. Assume further that core material exhibits under shock compression the same linear relation between shock wave velocity u_s and particle velocity u_p that is observed in many substances: $u_s = S u_p + C_0$. Here C_0 and S are characteristic constants of the material, C_0 is the sound velocity at $P = 0$, and S is a dimensionless parameter related to the pressure derivative of the incompressibility. Stewart shows how, given values of the "Hugoniot parameters" ρ_0, C_0, and S and of γ_0 and η, it is possible to calculate the sound velocity and a corresponding seismic parameter ψ at any density. Stewart then attempts to determine the values of the Hugoniot parameters that will give the best fit, at all depths in the outer core, between the calculated and observed seismic parameters ψ and ϕ. Once a choice of best-fitting Hugoniot parameters has been made, the temperature and the adiabatic gradient at any depth can be calculated, provided a value is chosen for c_V, the specific heat at constant volume. It turns out, however, that partly because of the uncertainty affecting seismic velocities and ϕ, there is a wide range of acceptable values of the Hugoniot parameters. Estimated temperatures vary widely with assumed Hugoniot parameters and assumed density distribution in the core. Stewart, using experimental parameters for Fe-Si alloys, shows that the temperature at the core-mantle boundary changes from about 1000°K to more than 5000°K when the assumed silicon content of the core is reduced from 20 percent to 10 percent.

Although these calculations are not directly applicable to the core, which probably does not consist of Fe-Si, they do serve to point out that whatever the core happens to consist of, its temperature cannot presently be calculated with any degree of accuracy, partly because of the uncertainty that affects the density and, to a lesser extent, the seismic velocity. The fact is that a change in temperature of 1000° probably changes density in the core by less than 1 percent. The same remark applies to the adiabatic gradient, which cannot be determined from seismic data to better than within a factor of 2 or more.

An entirely different approach is taken by Bukowinski (1977) to determine the temperature in the solid inner core, assumed to consist of pure iron. A quantum-mechanical electronic band structure calculation is made for γ (fcc) iron. The equation of state is taken to be of the Mie-Grueneisen type, with an appropriate term to represent the contribution to the thermal pressure from conduction electrons. Bukowinski then seeks the temperature at which his theoretical equation of state fits the geophysical data (P, ρ, ϕ) for the inner core. Five different proposed density distributions are examined, for which the temperature T_i at the inner core–outer core boundary ranges from a low of 4400° ± 40°K to a high of 8900° ± 100°K. Values of γ range from 0.41 ± 0.1 to 3.93 ± 0.1. Bukowinski's preferred solution, corresponding to the PEM model of Dziewonski *et al.* (1975), has $T_i = 5450° ± 65°K$ and $\gamma = 1.87 ± 0.1$. Surprisingly, the inner core turns out not to be isothermal, the temperature at the center being 234° higher than T_i; this implies heat sources in the inner core, which we shall consider later (Chapter 4).

Bukowinski ignores the possibility that the stable phase of iron under inner core conditions might be the ϵ (hcp) phase rather than the γ (fcc) phase he exclusively considers (Liu, 1975a). The density of the ϵ phase exceeds that of the γ phase by a few percent, enough perhaps to make a noticeable difference in the core temperature.

54 ENERGETICS OF THE EARTH

TABLE 3-1 Predicted Density with Depth

Depth (km)	Density (g/cm³)	
	Model CAL 5I1A[a]	Model PEM[b]
2885 (top of outer core)	9.95	9.909
5155 (bottom of outer core)	12.34	12.139
5155 (top of inner core)	13.11	12.794
6371 (center of earth)	13.35	13.01

[a] From Bolt and Uhrhammer (1975)
[b] From Dziewonski et al. (1975)

Clearly an accurate determination of temperature in the inner core requires a very precise knowledge of its density. Permissible models of the earth (i.e., models that satisfy constraints on total mass, moment of inertia, surface waves and body waves, and eigenvibrations) differ by notable amounts, as can be seen from Table 3-1. Although densities in these two models are almost identical at the top of the outer core, they differ by 3 percent at the top of the inner core and by 2.6 percent at the center.

We conclude, then, that partly because of uncertainty concerning the density and seismic velocity distributions in the core, it remains impossible to assess temperatures in the core from equations of state and geophysical data to better than within a few hundred or perhaps even a few thousand degrees.

THE ADIABATIC GRADIENT

A second method for determining temperatures in the core starts from the supposedly known temperature at some point (e.g., the core-mantle boundary) and assumes an adiabatic (isentropic) distribution. The adiabatic gradient is

$$\frac{dT}{dr} = -\frac{g\gamma T}{\phi}, \qquad (3.13)$$

where $\gamma = \alpha K_s/\rho c_P$ is the Grueneisen thermal ratio, $\phi = v_p^2$ is the seismic parameter determined from the sound velocity v_p, and g is the acceleration of gravity. All three quantities are functions of r. The temperature at radius r is then

$$T_r = T_0 \exp\left[\int_{r_0}^{r} \frac{-g\gamma}{\phi} dr\right], \qquad (3.14)$$

where T_0 is the reference temperature at $r = r_0$. $\phi(r)$ is known from seismic observations, and $g(r)$ may be calculated for any density model of the earth. It remains to calculate γ.

A recent example of this method of determining the temperature may be found in Stacey (1977), whose calculations illustrate the many pitfalls into which one is likely to fall and to which Knopoff and Shapiro (1969) drew attention. After making a number of dubious approximations, Stacey finds that γ in the outer core varies from 1.26 at the bottom to 1.42 at the top. These numbers fall in a plausible range, but so would numbers 50 percent bigger or smaller.

A surprising feature of many of the papers on the outer core's Grueneisen ratio is their insistence of referring it to the experimental values γ_0 of pure solid iron at normal pressure and temperature. The outer core is, of course, liquid, a fact that should throw at least some doubt on the validity of applying to it results straight out of crystal lattice dynamics. There seems to be a secret hope that properties of liquid iron might not, after all, be very different from those of solid iron. But this does not seem to be the case. At the melting point (1809°K) at normal pressure, the thermal expansion of solid iron is 69.5×10^{-6}/deg, while that of liquid iron is 119.2×10^{-6}/deg (Kirshenbaum and Cahill, 1962). The Grueneisen ratio of liquid iron at the melting point can be calculated from the sound velocity c since $\gamma = \alpha K_s/\rho c_P = \alpha c^2/c_P$. The value of c is 4.4 km/s according to Filipov *et al.* (1966), and 3.93 km/s according to Kurz and Lux (1969). The former value gives $\gamma = 2.8$,* the latter gives $\gamma = 2.23$; both values are much larger than the value for solid iron at room temperature (1.59). Whether the difference between the values for liquid and solid will increase or decrease with increasing pressure or density is not known. It is noteworthy that whereas γ for solids usually decreases with increasing density, the opposite is true for water and mercury (Knopoff and Shapiro, 1969).

However that may be, an average value $\bar{\gamma} = 1.3$ is not impossible. In the outer core, g varies from about 10 m/s² at the top to about 4.5 m/s² at the bottom, and ϕ varies from 66 km²/s² to about 100 km²/s². Taking averages, $\bar{g} \cong 7$ m/s² and $\bar{\phi} = 85$ km²/s², gives 1.275 for the ratio of the temperature at the inner core boundary (ICB) to the temperature at the outer core boundary (CMB). If the latter is taken to be 4500°K, the former is $\approx 5740°K$. The adiabatic gradient is $\approx 1°/km$ at the top, and 0.3°/km at the bottom. The temperature of 5740°K at the ICB may be compared to Bukowinski's (1977) estimate of $5,450° \pm 65°K$ referred to above. The general agreement between the two figures should not,

---

*Using $c_P = 11.0$ cal/mol deg (Anderson and Hultgren, 1962).

however, be considered to be much more than a coincidence, considering the amount of guessing that goes into the choice of $\bar{\gamma}$.

THE TEMPERATURE AT THE INNER CORE–OUTER CORE BOUNDARY

Since the inner core is solid and presumably consists of iron (plus, possibly, some nickel and small amounts of a few other less abundant elements), while the outer core, consisting mostly but not entirely of iron, is liquid, it is rather natural to assume that the temperature at the boundary must be such that solid iron is in equilibrium with its melt, at the prevailing pressure.

The temperature at which a pure substance is in equilibrium with its melt is called the melting point T_m. It depends only on pressure, as

$$\frac{dT_m}{dP} = \frac{\Delta V}{\Delta S}, \tag{3.15}$$

where $\Delta V = V^l - V^s$ is the difference in volume between liquid and solid, and ΔS is the corresponding difference in entropy. In a multicomponent system, the temperature at which a solid phase is in equilibrium with liquid depends not only on pressure but also on the composition of the liquid. For instance, in a system consisting of iron and x percent of sulfur at $P = 1$ bar, solid iron can be in equilibrium with liquid at any temperature between 1809°K (for $x = 0$, pure iron) to 1261°K (for $x = 31.4$). Since addition to a pure substance of any component that is soluble in the melt necessarily lowers the melting point of the pure substance, the melting point of pure iron at the pressure of the inner core boundary sets an upper limit to the temperature there.

THE MELTING POINT OF IRON

The pressure dependence of the melting point of iron has been the subject of much debate in recent years (for good summaries see Jacobs, 1975, and Boschi, 1975). Some measurements of doubtful accuracy have been carried to 200 kbar. The most frequently cited data are those of Sterrett *et al.* (1965) up to 40 kbar. The problem is to extrapolate those results to the pressure (3.3 Mbar) of the inner core.

It is curious that we are still unable to explain why solids melt, or, for that matter, why the liquid state exists at all. All theories of melting are essentially empirical. A theory that has had much success is that of Lindemann. Lindemann supposes that melting occurs when the amplitude of thermal vibrations reaches a fraction δa of the lattice spacing a. In that state, at temperature T_m, the kinetic energy of vibration $3/2\, kT_m = 1/2\, m\, (\delta a)^2 \omega^2$, assuming that atoms of mass m vibrate harmonically with frequency ω; k is Boltzmann's constant. The effect of pressure on T_m will arise from its effect on a, which is proportional to $V^{1/3}$ (V is volume), and on ω. It is convenient here to introduce the Debye temperature $\theta = h\omega/k$ (h is Planck's constant), and recall that from Debye's theory $\theta \cong V^{1/3} v$, where v is the mean elastic velocity defined in Equation (3.9). Thus T_m is directly proportional to v^2, and the ratio of the square of the elastic velocity in the inner core to that of iron at $P = 1$ bar gives the ratio of the melting temperatures. Alder (1966) obtained in this manner a melting temperature of 7720°K.

Attempts have been made to refine the Lindemann law by reformulating it in less empirical terms. Boschi (1974), using a melting criterion by Ross, derives the relation

$$T_m = T_0 \left\{ 1 - \frac{n}{3}\frac{\Delta V}{V_0} + \frac{n}{6}\left(\frac{n}{3}+1\right)\left(\frac{\Delta V}{V_0}\right)^2 + \ldots \right\}, \qquad (3.16)$$

where T_m is the melting point at a pressure at which the volume of the solid is V, T_0 is the melting point in some reference state (e.g., $P = 0$) at which the volume is V_0, and $\Delta V = V_0 - V$. Here n is the exponent in the expression for the potential ψ of the repulsive interatomic force, assumed to be of the form $\psi \cong r^{-n}$, where r is the interatomic distance. For iron, $n \cong 8.4$, as determined from isothermal compression tests. The validity of (3.16) depends on how well the power law r^{-n} represents the interatomic potential; different expressions for that potential lead to widely different values of T_m when $\Delta V/V$ is large.

Kraut and Kennedy (1966) suggested that melting points can very generally be represented as

$$T_m = T_0 \left(1 + C\, \frac{\Delta V}{V_0}\right), \qquad (3.17)$$

where ΔV and V_0 have the same significance as above, and C is a constant that can be determined from the initial slope of the melting point curve. Higgins and Kennedy (1971) extrapolated the data of Sterrett et al. (1965) to 3.3 Mbar to obtain a melting point of 4250°C ($\approx$4500°K).* How the volume V of the solid at its melting point at 3.3

---

*But the initial slope of the melting curve of Sterrett et al. (2.85°C/kbar) does not agree with the slope (3.8°/kbar) calculated from measurements of ΔV and ΔS and the Clapeyron equation (3.15).

Mbar is determined is not said, though the authors do state that the density at 1.5 Mbar is assumed to be 11.4 g/cm³.

Although Equation (3.17) represents fairly well the melting point behavior of some substances at small compression, it is unlikely that it will apply to all substances, as Kraut and Kennedy (1966) had originally claimed. Validity of the simple linear relation is now claimed only for metals (Higgins and Kennedy, 1971). If (3.17) is valid it follows that

$$\frac{dT_m}{dP} = T_0 C\beta \frac{V}{V_0},$$

where β is compressibility. Since β and V are necessarily positive quantities, dT_m/dP cannot change sign and T_m can go through neither a maximum nor a minimum. Yet there are many metallic elements (cesium, barium, gallium, silicon, antimony, bismuth, selenium, tellurium, cerium, uranium) that do just that; for lithium and potassium the slope of the melting point almost goes to zero, even though β does not. (For a compilation of phase diagrams at high pressures, see Cannon, 1974.) It is true that many of the observed changes in slope of the melting curve (T_m, P) occur in conjunction with a change of phase of the solid, yet the maximum melting point for barium occurs between 10 kbar and 20 kbar, whereas the nearest phase change occurs along the melting curve only above 60 kbar. The melting point curve of selenium has a maximum near 50 kbar, but no reported phase change (note that selenium is mentioned by Higgins and Kennedy as an example of an element satisfying the linear law). Attributing changes in slope to phase changes does not help much for iron, since between the pressure range of the observations by Sterrett *et al.* and of the inner core, iron undergoes at least two phase changes: $\delta \to \gamma$ and $\gamma \to \epsilon$ (Liu, 1975a). On the basis of the γ-ϵ transition, Birch (1972) estimated that even if the linear extrapolation were correct, the melting point at core pressures might have to be raised by some 700°.

A comparison of (3.17) and (3.16) makes it clear that the linear relation (3.17) is an approximation that can be valid only in the limit of small $\Delta V/V_0$ or if n is small, as it would be for soft metals with high compressibility. Quite apart from the effect of phase changes, the linear law for iron almost certainly breaks down, perhaps seriously, at pressures in the megabar range. Boschi (1974), using Equation (3.16), finds a melting point at 3.2 Mbar of 6600°C, as against Higgins and Kennedy's 4250°C. Note, however, that the figure arrived at by Boschi strongly depends on his choice for the repulsive potential. The volume dependence of the melting point, as with γ, to which the melting point is related in Lindemann's theory, is a sensitive function of the interatomic potential.

Most theories of melting suffer from the defect that they attempt to predict the melting point by considering only properties of the solid phase. Since the melting point is, by definition, the temperature at which the free energy of the solid equals that of the liquid, it should reflect properties of both the solid and liquid phases. Leppaluoto (1972a,b) has attempted to do this, using Eyring's significant-structure theory to predict properties of liquid iron. It will be recalled that this theory represents a liquid as consisting on the one hand of "solidlike" atoms with vibrational degrees of freedom as in the solid, and on the other hand, of "gaslike" atoms that have acquired translational degrees of freedom that allow them to jump into unoccupied sites or "holes," the number of which determines the difference in volume between liquid and solid. A partition function is then written to represent the "solidlike" and "gaslike" structures and any other structures (magnetic, diatomic, etc.) that may be significant.

Leppaluoto's work has been dismissed by Boschi (1974) as "lacking credibility" because an early attempt by Tuerpe and Keeler (1967) to apply significant-structure theory had led to anomalous results. Leppaluoto's work was designed, however, to circumvent these difficulties. He first writes down a partition function for liquid iron at $P = 1$ bar in which the parameters are chosen so as to give the correct ΔV and ΔS of melting and the correct Gibbs free energy change $\Delta G = 0$ at the melting point (1809°K). The lattice or "solidlike" properties of solid iron at high temperature are calculated in the Einstein approximation. The partition function so designed predicts within 5 percent the thermal expansion and compressibility of liquid iron, a slightly high (20 percent) specific heat, and a good temperature dependence of viscosity.

The melting point at high pressure is then determined as the temperature at which the free energies of solid and liquid are equal. Properties of the solid are determined from shock wave data. The calculations involve a quantity ΔV^*, an activation volume, which measures the pressure dependence of the free energy of activation a gaslike particle must have to move into a hole; to occupy a hole a molecule of liquid must do work $P \Delta V^*$ against the external pressure P. ΔV^* is similar, but not exactly equal, to the activation volume for self-diffusion, and is not well determined. Leppaluoto finds, however, that agreement between his calculated melting point and the melting point calculated from the Clapeyron equation requires ΔV^* to be constrained between 0.07 and

60 ENERGETICS OF THE EARTH

0.18 cm³/mol at pressures up to 3.3 Mbar. The lower limit for ΔV^* gives $T_m = 7000°K$ at 3.3 Mbar; the upper limit gives 9500°K. For $\Delta V^* = 0$, which leads to a melting temperature inconsistent with the Clapeyron equation, the melting point at 3.3 Mbar is close to the value predicted by Higgins and Kennedy (1971). There is also some uncertainty in the melting point arising from the fact that shock wave data presumably refer to the ϵ (hcp) phase of iron, not to the δ (bcc) phase present at the melting point at $P = 1$ bar. Taking this into account, Leppaluoto suggests $T_m = 7500° \pm 2000°K$ for iron at 3.3 Mbar. Alder's (1966) and Boschi's (1974) results fall within these limits.

MELTING IN THE FE-S SYSTEM

As mentioned earlier, the melting point of pure iron at 3.3 Mbar ($\approx 7500°K$) is an upper bound on the temperature at the inner core–outer core boundary, since the outer core presumably does not consist of pure iron. The temperature at which solid iron is in equilibrium with a multicomponent melt will always be less than the melting point of pure iron by an amount that depends on the nature and proportion of the other components. Oxygen, for instance, has very little effect on the melting point of iron until its proportion reaches about 23 percent by weight; but even then the lowering of the melting point is less than about 150°. On the other hand, addition of sulfur may lower the melting point of iron by more than 500°. The lowest temperature at which solid iron can be in equilibrium with an Fe-S melt is 988°C at $P = 1$ bar (eutectic temperature); the melt then contains 31.4 percent of sulfur. At higher sulfur content, the solid phase in equilibrium with the melt (above 988°C) is FeS. The phase diagram for the Fe-S system at 1 bar is shown in Figure 3–6.

Just as the melting point of pure iron is an upper bound to the temperature at the inner core boundary, so the eutectic temperature is a lower bound. The effect of pressure on the eutectic temperature T_e has been measured at 30 kbar by Brett and Bell (1969) and at from 30 to 100 kbar by Usselman (1975a). From an initial value of 988°C at $P = 1$ bar, T_e first rises very slowly, to 998° ± 5°C at 55 kbar, then much more rapidly, to 1190°C at 100 kbar. The break in slope, which occurs near 52 kbar, is presumably connected with a phase change in solid FeS, or with the substitution of FeS_2 for FeS as the stable form of iron sulfide; this is possible because the reaction

$$2FeS = FeS_2 + Fe$$

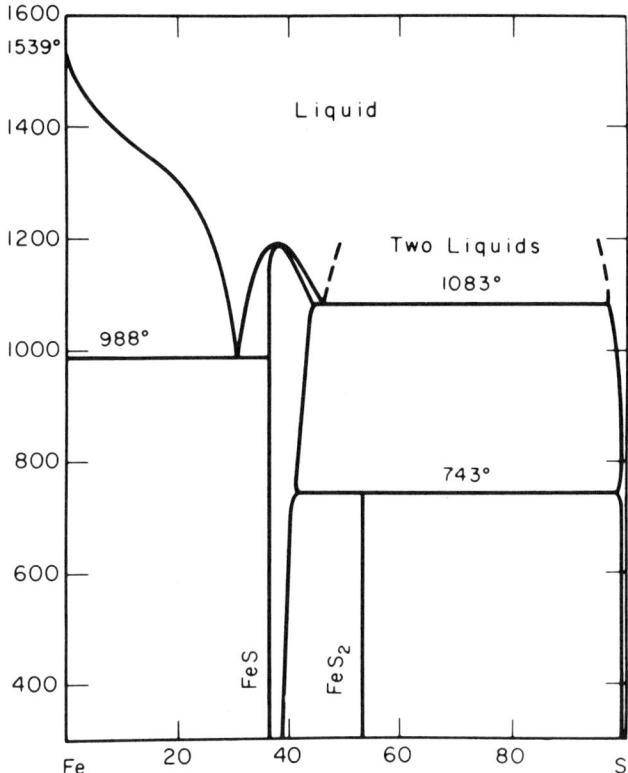

FIGURE 3–6 Phase diagram for the Fe-S system at ordinary pressure.

has a large negative volume change $\Delta V = -5.31$ cm^3 at ordinary pressure. As the pressure rises, the sulfur content of the eutectic mixture first decreases from 31.4 percent at $P = 1$ bar to 24 percent at 55 kbar, and remains essentially constant from there on.

Attempts to extrapolate the eutectic temperature to core pressures have been made by Usselman (1975b) and Stacey (1977). Usselman uses the Kraut-Kennedy linear extrapolation, but since the compressibility of solid mixtures of Fe and FeS is very poorly known, the volume of the solid phases at 3.3 Mbar has to be guessed. Stacey uses a form of Lindemann's theory that requires knowledge of γ, the Grueneisen ratio for the solid phase; since this is not known, Stacey uses the value he derived for the liquid outer core. Both calculations ignore the γ-ϵ transition in iron, which is likely to change the slope

dT_e/dP just as the phase transition in FeS steepens it at 52 kbar. At 3.3 Mbar, T_e is between 3750° and 4050°K according to Usselman, and 4168°K according to Stacey.

The effect of pressure on the eutectic is formally described by the relations (Prigogine and Defay, 1954, p. 365)

$$\frac{dT_e}{dP} = T \frac{x_1 \Delta v_1 + x_2 \Delta v_2}{x_1 \Delta h_1 + x_2 \Delta h_2} = T \frac{v^l - v^s}{s^l - s^s}, \qquad (3.18)$$

$$\frac{dx_2}{dP} = -x_1 \frac{\Delta h_1 \Delta v_2 - \Delta h_2 \Delta v_1}{RT(x_1 \Delta h_1 + x_2 \Delta h_2)(\partial (\ln x_2 \gamma_2)/\partial x_2)}, \qquad (3.19)$$

where x_1 and x_2 are, respectively, the mole fractions of components 1 and 2 at eutectic composition, γ_2 is the activity coefficient of component 2 in the melt, and

$$\Delta h_i = \bar{h}_i^l - h_i^s, \qquad \Delta v_i = \bar{v}_i^l - v_i^s, \qquad (i = 1, 2),$$

where $\bar{h}_i$ and $\bar{v}_i$ are, respectively, the partial molar enthalpy and volume of component i, and superscripts l and s refer to liquid phase and solid phase, respectively. It is important to note that although v_i^s at any pressure could be calculated from an adequate equation of state, the partial molar volume $\bar{v}^l$ cannot because it depends on volume changes that occur, at constant P and T, when Fe and FeS mix. If the FeS melt were a perfect solution, with no volume change upon mixing, $\bar{v}_i^l = v_i^{l_0}$, the molar volume of pure liquid i. The same remark applies to enthalpy, which would be calculable only if there were not heat of mixing, as in a perfect solution. That FeS liquids are not perfect solutions is shown experimentally by the very fact that the slope of the eutectic temperature curve is essentially zero up to 55 kbar. This requires the numerator of (3.18) to be zero, and since x_1 and x_2 are both positive quantities, either Δv_1 or Δv_2 must be smaller than 0. The volume of the eutectic liquid is less than the sum of the volumes of its two pure liquid components, so contraction occurs upon mixing.

Because of the markedly nonideal nature of the FeS system, it is in fact impossible to predict what its phase diagram will look like at high pressure. The solid phase FeS may become unstable with respect to FeS_2 or a high-pressure phase of it. Pyrite (FeS_2) melts incongruently at 743°C at $P = 1$ bar. The very existence of a eutectic between Fe and FeS (or FeS_2) may be in doubt. As noted by Kullerud (1970), most sulfur-metal systems exhibit liquid immiscibility, i.e., the existence of not just one but of two coexisting liquid phases. The systems Cu-S,

Pb-S, and Hg-S have in fact two immiscibility ranges, one of which occurs at low sulfur concentration. In the Cu-S system at $P = 1$ bar, for instance, there is a eutectic with only 0.77 percent sulfur at 1067°C, barely below the melting point of pure copper (1083°C). At 1105°C, a liquid with 1.5 percent sulfur is in equilibrium with a liquid containing 19.8 percent sulfur, the composition of which is close to that of Cu_2S, which melts congruently at 1129°C. The second immiscibility gap occurs at much higher concentrations of sulfur. This led Verhoogen (1973) to suggest that immiscibility in the Fe-S system could perhaps account for the properties of the lower few hundred kilometers of the outer core, which were once thought to form a separate layer (Bullen's layer F) with properties (e.g., density) different from those of the rest of the outer core (Bolt, 1972). It now seems that the properties of layer F are not sufficiently different to warrant its recognition as a separate entity, even though the decrease in slope of the seismic velocity versus depth curve has not been satisfactorily explained.

Quite apart from the highly questionable propriety of applying, as Usselman and Stacey do, to a eutectic questionable theories of melting in one-component systems, it would seem that, from the very nature of the Fe-S system (or of the Fe-O system, for that matter), the determination of its eutectic temperatures at very high pressure is even more. uncertain than determining the melting point of pure iron. For all it is worth, one could perhaps venture to guess that since the slope (4.2°/kbar) of the eutectic temperature curve above 55 kbar is slightly steeper than the initial slope of the melting curve of iron (3.8°/kbar), pressure would tend to reduce the difference between the melting point of iron and the eutectic temperature. But this extrapolation ignores possible differences in compressibility between liquid iron and Fe-S melts that could reduce the slope of the melting curve more quickly for the eutectic liquid.

In spite of all this uncertainty, Stacey (1977) has recently proposed a temperature distribution for the whole earth that is anchored to a temperature of 4168°K at the boundary of the inner core. That temperature of 4168°K is, it will be recalled, Stacey's estimate of the eutectic temperature at inner core pressure, the basic assumption being that the liquid outer core and solid inner core have the same (eutectic) composition. This assumption is almost certainly false, for two reasons. The first reason is that the density of the inner core is compatible with it consisting of iron (or iron plus a small amount of nickel), but there is no evidence to support the contention that an Fe-FeS solid eutectic mixture (zero pressure density ≈ 5.5 g/cm^3) could attain a density of about 13 g/cm^3 at 3.3 Mbar (see Table 3–1). The second reason is that, if the

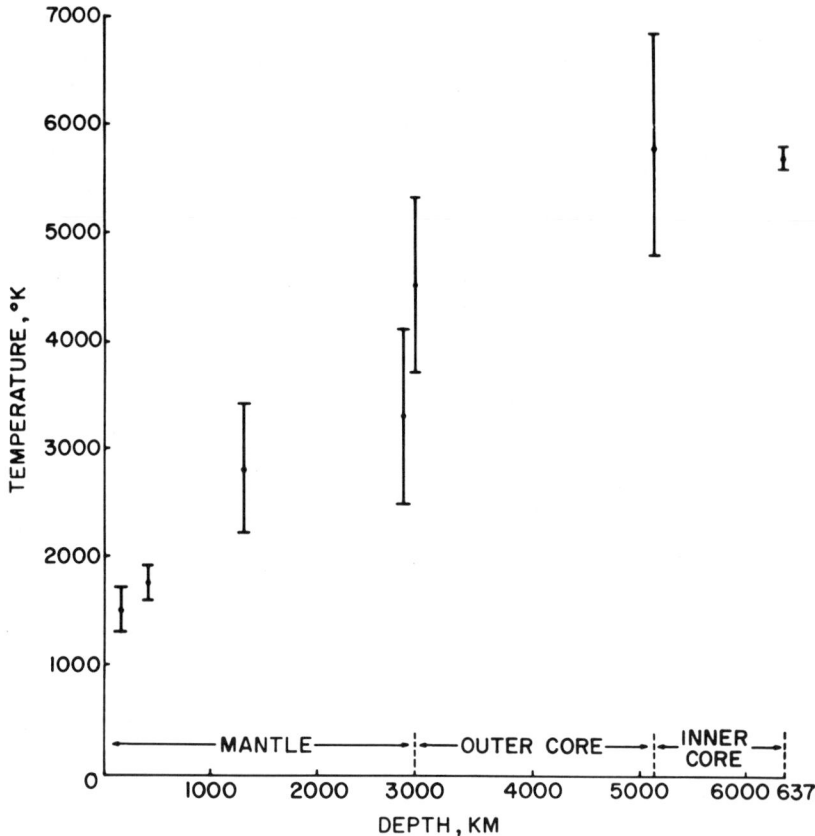

FIGURE 3-7 Temperatures in the earth, estimated by methods explained in the text. Points shown refer to, from left to right: (1) xenoliths in kimberlite; (2) the olivine-spinel transition at 400 km; (3) the lower mantle at 1300 km, from Wang (1972); (4) the lower mantle at 2800 km, also from Wang; (5) the core-mantle boundary; (6) the inner core boundary; and (7) the center of the earth, from Bukowinski (1977). Estimates of uncertainties are somewhat arbitrary, as they include many diverse factors, such as regional variations at point (1). The error bar at point (7) (center of the earth) reflects only the uncertainty inherent in Bukowinski's calculations, which assume that the inner core density is exactly that given by the PEM model of Dziewonski et al. (1975). Bukowinski estimates that an error of 1 percent in the density adds about 500° to the uncertainty on the temperature.

inner and outer cores have the same eutectic composition, the density jump of 0.6–0.8 g/cm³ at their interface (Table 3–1) necessarily represents solely the effect of melting at constant composition. When introduced in Equation (3.18) for the pressure dependence of the eutectic temperature, it leads to a very high value, of the order of several degrees per kilobar, for the slope of the eutectic line, which is inconsistent with Stacey's value of the eutectic temperature itself. It seems safe to conclude that the sulfur content of the outer core is less than that of the eutectic and that, accordingly, the temperature T_i at the boundary of the inner core is higher than the eutectic temperature, whatever that temperature may be.

In summary, then, all that can be said about T_i is that it is less than the melting point of pure iron at 3.3 Mbar (7500° ± 2000°C), but how much less is not known. Bukowinski's estimate of 5450°K is not excluded; but recall that it is calculated from the properties of γ iron, which may be irrelevant to the inner core, and for a particular density model that assigns to the inner core a somewhat lower density than other models do (Table 3–1).

All in all, it seems likely that temperatures in the core are in the broad range between 4500° ± 800°K at the core-mantle boundary and 6000° ± 500°K at the center. Those temperatures are somewhat higher than the estimated initial temperatures calculated from accretion theory (Hanks and Anderson, 1969), which are particularly low for the core because the rate of release of gravitational energy is necessarily low at the beginning of accretion, when the gravitational pull of the accreting body is still quite small. Our temperatures suggest that the core has heated up in the course of time. Several sources immediately suggest themselves (Chapter 2). If the core now contains 0.1 percent of potassium, the total heat generated by it through 4.5 billion years amounts to 10^{30} J, enough to heat up the core by some 700°; if one half of the gravitational energy of separation of the core (1×10^{31} J, according to Flasar and Birch, 1973) were transformed into heat in the core, it would raise the core's temperature by some 3500°.

SUMMARY

In Figure 3–7 we have plotted as a function of depth the possible temperatures, or temperature ranges, deduced from the considerations outlined above. These temperatures can be summarized as follows:

1. From temperatures of incipient melting at base of lithosphere:

    oceanic lithosphere 50 km thick: 1075° ± 225°C
    oceanic lithosphere 100 km thick: 1150° ± 250°C
    continental lithosphere 200 km thick: <1200°C

2. From kimberlite nodules:

$$\text{at 100 km: } 1000° \pm 200°C$$
$$\text{at 150 km: } 1200° \pm 200°C$$

Clearly it is not possible to draw a single solution through these points. Regional variations are considerable. A (possibly meaningless) average temperature gradient in the first 100 km might be 10–12°/km.

3. From olivine-spinel transition:

$$\text{at 400 km: } 1450° \pm 150°C$$

4. Mean adiabatic (isentropic) gradient between 100–700 km:

$1.5° \pm 0.5°$/km (including entropy change due to phase transitions)

This gives:

$$\text{at 700 km: } 2000° \pm 350°C = 2300° \pm 350°K$$

5. Below 700 km, the isentropic gradient falls to 0.3°–0.4°/km. The following temperatures are those calculated by Wang (1972), assuming adiabaticity:

$$\text{at 1300 km: } 2800° \pm 800°K$$
$$\text{at 2800 km: } 3300° \pm 800°K$$

6. In boundary layer D'', the gradient is estimated at 10°–12°/km. The temperature at the core boundary is then $4500° \pm 800°K$. The melting point of pure iron at the pressure of the core-mantle boundary is $5000° \pm 500°K$ (Leppaluoto, 1972a,b).

7. Guessing that an average value of γ suitable to the outer core is about 1.3, the temperature at the ICB comes out as 5740°K, with an uncertainty possibly as large as 1000°. Bukowinski's (1977) estimate is $5450° \pm 60°K$. Bukowinski's temperature at the center of the earth, plotted in Figure 3–7, is $5684° \pm 65°K$, a value predicated on the assumptions (1) that the inner core consists of iron in its γ (fcc) form, and (2) that the density of the inner core is that given by the so-called PEM model. Bukowinski estimates that a change in density of 1 percent from the PEM value entails a change in temperature of about 500°.

4 Dynamics of the Core

We now turn to a closer study of the outer core. Our purpose is to discover how much energy is generated in it, how much heat crosses into the mantle, whether this amount of heat is consistent with our earlier consideration of the lower mantle and of layer D'' in particular, and whether it is sufficient to drive whole-mantle convection.

Our principal clue is the geomagnetic field. It is now generally agreed that this field is generated in the outer core by motion (flow) of the electrically conducting fluid (molten iron and sulfur) that forms the outer core. Since precession seems unlikely to be the main cause of the flow (Rochester *et al.*, 1975; Loper, 1975), convection is required. Convection can be either thermal or chemical. Thermal convection requires a source of heat. In chemical convection, the differences in density that cause the motion are due to differences in chemical composition (e.g., a difference in sulfur content). It is indeed conceivable that slow cooling of the whole core might lead to crystallization of iron, which accumulates to form a growing inner core; removal of iron from an Fe-S liquid leaves a liquid richer in sulfur and presumably lighter than the rest of the outer core; the lighter liquid rises by buoyancy. Both the settling of the solid iron and the rising of the sulfur-enriched fluid release gravitational energy; hence the name gravitationally powered dynamo given to generation of the magnetic field by this process, which has long been advocated by Braginsky (1963), and more recently by Gubbins (1977), by Loper (1978), and by Loper and Roberts (1978). The name is perhaps not quite appropriate, since a thermally driven convecting system is also powered gravitationally; in addition, it should be remembered that the assumed crystallization of iron and formation of a light liquid enriched in sulfur result from cooling and are therefore also thermal phenomena.

STABILITY OF THE CORE

Whether convection is thermal or chemical, the temperature gradient in the outer core cannot, on the average, depart very much from being adiabatic. If it could somehow be shown that the temperature gradient is less than adiabatic, it would follow that convection does not occur because the core is stably stratified.

One of the few things that seem certain for the core is that since the inner core is solid and the outer core liquid, the temperature curve must be below the melting point curve in the inner core and above it in the outer core; at the inner core boundary (ICB), where $r = r_i$, the temperature gradient must be less, in absolute value, than the melting point gradient. This is sketched out in Figure 4–1. If the core is unstable and convecting, the temperature gradient must be steeper than the adiabat and so also must the melting point curve. Higgins and Kennedy (1971) have argued that this is not the case; but their conclusion is based on an estimate of the melting point of iron at the inner core boundary that is almost certainly too low (see Chapter 3). As also mentioned in Chapter 3, the adiabatic gradient in the outer core is uncertain by a factor of perhaps as much as 2. Thus no firm conclusion can be reached on those grounds.

Is there observational evidence to the effect that the core is stably stratified or not? Olson (1977) has considered the problem of the internal oscillations of a body consisting of a uniform solid elastic mantle and a solid inner core bounding a stratified, rotating, inviscid, fluid outer core. The interaction of buoyancy and rotation results in two types of waves: (1) internal gravity waves that exist if $N^2 > 0$ (N is the Brunt-Väisälä frequency, which describes the density stratification; $N > 0$ in a stable core); and (2) inertial oscillations that exist if $N^2 < 4\Omega^2$ (Ω is the angular velocity of rotation). For a model with a density stratification similar to that proposed by Higgins and Kennedy, the internal gravity waves have eigenperiods of at least 8 hours. A model with unstable stratification admits no gravity waves but admits inertial oscillations whose eigenperiods have a lower bound of 12 hours. There is unfortunately almost no observational evidence of long-period terrestrial oscillations; in any event, as Olson points out, their amplitude is confined predominantly to the outer core, so that their detection

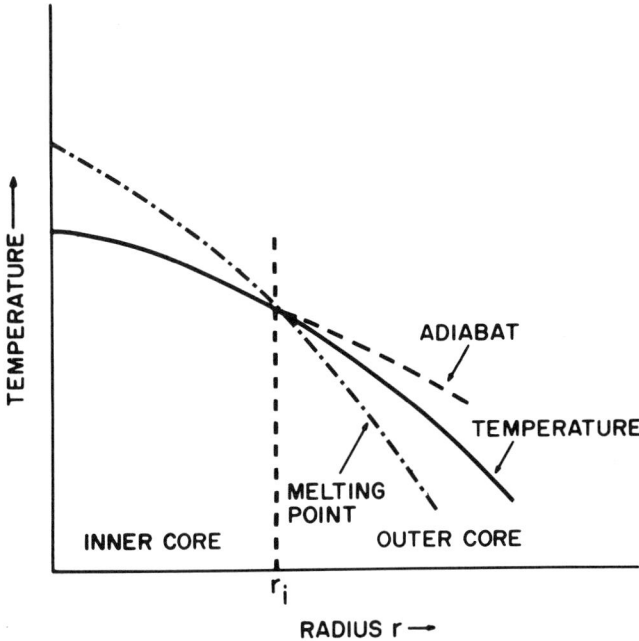

FIGURE 4-1 Temperature in the core. The temperature must lie *below* the melting curve in the solid inner core, and above the melting curve in the liquid outer core. For convection to occur throughout the outer core, the temperature in it must lie below the adiabat through the inner core boundary.

on the surface would be difficult. Since, furthermore, oscillations in the real core are likely to be hydromagnetic rather than mechanical, observed periods of free oscillations of the earth cannot as yet be used to discriminate between stable and unstable core.

In a gravitational field, a two-component fluid that is not stirred will develop a compositional gradient, because the lighter constituent (sulfur in the case of the core) will tend to concentrate at the top (Guggenheim, 1933, pp. 103–159). Since sound velocity depends on composition as well as on pressure and temperature, the sound velocity profile through the outer core could be used, at least in principle and given an adequate equation of state, to detect any such compositional gradient. None has been clearly revealed to date. Too little is known, however, of the diffusion coefficient of sulfur in an iron-sulfur melt under core conditions to predict whether the sulfur distribution could reach its equilibrium value within the few billion years of the core's existence.

Since, on the one hand, there is no observational evidence that the core is stably stratified, and since on the other hand there is a geomagnetic field, the generation of which does seem to require flow, we shall assume henceforth that the outer core is indeed convecting, with the implication that the temperature curve through the outer core is only very slightly steeper than the adiabat through the ICB and lies above the melting point curve. This will require the temperature at the inner core boundary to be on the high side of the estimates given in Chapter 3.

CAUSES FOR CONVECTION

Conceivably, the outer core could be convecting for two different reasons:

1. The outer core contains distributed heat sources, for instance radioactive potassium (see Chapter 2).
2. Cooling from the top. Convection can occur in a fluid cooled from the top as well as in a fluid heated from below, or heated internally. The mantle, losing heat, as it does at the surface of the earth, may be cooling, with the temperature at the core-mantle boundary (CMB) slowly decreasing with time.

Two further consequences of the cooling hypothesis are interesting:

1. Since the temperature at the boundary of the inner core equals the liquidus temperature (melting point), which increases downward with increasing pressure, cooling moves the ICB outward. Crystallization of iron releases some latent heat of melting, which slows down the cooling and increases the average temperature gradient through the outer core. The release of latent heat is, in effect, equivalent to a heat source. The outer core is thus cooled from the top *and* heated from below (Verhoogen, 1961).
2. Growth of the inner core releases gravitational energy. Crystallization of iron enriches the remaining liquid in sulfur, prompting what was described earlier as chemical convection.

ENERGY REQUIREMENTS OF THE GEOMAGNETIC DYNAMO

Associated with a magnetic field **B** there is an energy density $B^2/8\pi\mu$. In the electromagnetic units we shall be using, $\mu = 1$ in the core, which is assumed not to be ferromagnetic.

The electric current density $\mathbf{j} = (1/4\pi\mu)$ curl $\mathbf{B}$ leads to ohmic dissipation (=Joule heating) at the rate j^2/σ per unit volume. Here σ is the electrical conductivity of the core, assumed to be uniform. The value of σ generally accepted at the moment is 5×10^{-8} s/cm² = 5×10^5 mho/m (Gubbins, 1976); it used to be 3×10^{-6}. Reasons for the change are not clear; in any event the value of σ is uncertain, perhaps by as much as a factor of 10. This is not surprising, considering that neither composition nor temperature are exactly known.

Ohmic dissipation leads to gradual decay of the electric currents and of their magnetic field. In the absence of any source to maintain it, the field decays with time according to the equation

$$\frac{\partial \mathbf{B}}{\partial t} = \frac{1}{4\pi\mu\sigma} \nabla^2 \mathbf{B}, \tag{4.1}$$

which shows that a characteristic decay time τ is of the order of $\mu\sigma L^2$, where L is a characteristic length such that $\nabla^2 \mathbf{B} \cong B/L^2$. For the core, if L is taken to be 3×10^8 cm (approximately the radius of the core), $\tau \cong 4.5 \times 10^{11}$ s = 15,000 yr. Since paleomagnetic observations indicate that the strength of the earth's field 2 billion years ago oscillated within roughly the same limits as the field of the past few thousand years, it follows that the core must be producing magnetic energy at least at the rate E_m/τ, where

$$E_m = \int \frac{B^2}{8\pi\mu} dV$$

is the total magnetic energy, integration being over all space. Calculation of E_m is conveniently split into two parts. First we calculate the energy E_1 outside the core, where the field is a potential field characterized by the usual gaussian coefficients that describe the field at the surface of the earth. The first term in the expansion (dipole field) gives 5×10^{25} ergs. Other harmonic terms get progressively smaller as their order increases. The grand total is unlikely to be greater than about 5×10^{26} ergs. Taking $\tau \cong 5 \times 10^{11}$ s, as above, the dissipation rate is $\approx 10^{15}$ ergs/s = 10^8 W. Gubbins (1975), by a different method, calculates a dissipation rate ϕ for the poloidal field, $\phi \leq 1.5 \times 10^8$ W.

Inside the core, the situation is more complicated. In addition to the poloidal field B_p that emerges from the core as the potential field we measure at the surface, there may be a toroidal field B_ϕ, the field lines of which remain within the core* and which cannot therefore be measured directly. The existence of this field, which only has a component B_ϕ in the azimuthal direction, is made very likely by the ease with which it can be induced by any differential rotation within the core. If two parts of the fluid core rotate at different rates with angular velocities that depend on r or θ (r is the radial coordinate, θ is colatitude), the field lines of any poloidal field will be stretched and dragged by the differential rotation and wrapped around the rotation axis, forming a toroidal field that must necessarily vanish on the surface of the conducting fluid.

Since the strength of the toroidal field cannot be measured, it must be guessed. Guessing customarily proceeds along the following lines. The magnetohydrodynamic induction equation is

$$\frac{\partial \mathbf{B}}{\partial t} = \eta \nabla^2 \mathbf{B} + \text{curl}\,(\mathbf{u} \times \mathbf{B}), \tag{4.2}$$

where $\mathbf{u}$ is the velocity of the fluid motion and $\eta = 1/4\pi\mu\sigma$ is the magnetic diffusivity. In spherical coordinates (r, θ, ϕ) the ϕ component of (4.2) is, for an axisymmetric field:

$$\frac{\partial B_\phi}{\partial t} - \eta \left(\nabla^2 - \frac{1}{r^2 \sin^2 \theta} \right) B_\phi = B_r \left(\frac{\partial u_\phi}{\partial r} - \frac{u_\phi}{r} \right). \tag{4.3}$$

In the steady state, $\partial B_\phi/\partial t = 0$. It may be reasonably assumed that all remaining terms in equation (4.3) are of the same magnitude, so that, in particular,

$$\eta \frac{B_\phi}{r^2} \cong B_r \frac{u_\phi}{r}, \text{ or } B_\phi = B_r R_m, \tag{4.4}$$

where the Reynolds magnetic number R_m is defined as

$$R_m = \frac{u_\phi R}{\eta} = 4\pi\mu\sigma R u_\phi, \tag{4.5}$$

R being the radius of the core. It is also customary to suppose that u_ϕ may be about 4×10^{-2} cm/s. (The figure is obtained by interpreting the

---

*This assumes the lower mantle to be a perfect electric insulator, which it is not. The toroidal field of the core does probably leak into the lower mantle with much diminished intensity, but it still does not reach the earth's surface because of the very low conductivity of the upper mantle.

rate of westward drift of the secular variation, 0.2°/yr, to represent the differential rate of rotation of the outer layers of the core with respect to the rest of it and to the mantle.) Then, for $\eta = 2 \times 10^4 \text{cm}^2/\text{s}$, $R_m \cong 600$. Since B_r at the surface of the core is ≈ 4 G (extrapolated from its value, ≈ 0.5 G, at the earth's surface), B_ϕ might be as large as 2400 G.

The argument is quite speculative. Equation (4.4) is obtained by crudely equating terms on the right and left sides of (4.3); this procedure amounts to saying that because $100 - 99 = 2 - 1$, $100 \cong 2$ and $99 \cong 1$.

A different estimate of the total field B in the core is obtained by assuming a rough balance between the Coriolis and Lorentz forces. (The Lorentz force $\mathbf{j} \times \mathbf{B}$ is the force the magnetic field exerts on the fluid.) This gives

$$2\rho\Omega u = \frac{B^2}{4\pi\mu R}, \qquad (4.6)$$

since $\mathbf{j} = (1/4\pi\mu)$ curl $\mathbf{B}$ and curl $\mathbf{B} \cong B/R$. The density ρ of the core being about 11 g/cm³ and the mean angular velocity $\Omega = 7.29 \times 10^{-5}$/s, Equation (4.6), with $R = 3 \times 10^8$ cm and $u = 4 \times 10^{-2}$ cm/s as before, gives $B \cong 300$ G. The justification for equating Coriolis and Lorentz forces comes mostly, it seems, from Chandrasekhar's (1961) calculations on the onset of convection in a plane layer of viscous fluid that is rotating in the presence of a uniform magnetic field. Chandrasekhar showed indeed that under some special circumstances convection is most easily started (minimum critical Rayleigh number) when the two forces are approximately in balance (Acheson and Hide, 1973, p. 213). But this view has been criticized by Busse (1975b) and also by Gubbins (1976), who finds evidence against the existence of a toroidal field as large as 100 G (10 mT). Gubbins' argument is, however, a bit circular; he rejects a strong toroidal field because it leads to what he thinks is an unreasonably high rate of ohmic heating. Busse, on the other hand, reaches the conclusion that the toroidal field in the core is of the same order of magnitude as the poloidal field by solving the complete hydromagnetic problem (including the Lorentz force) in a cylindrical configuration that reproduces, he believes, the essential features of the core.

We are thus left guessing as to what the intensity of the magnetic field might be inside the core. If the average field in the core is 100 G (.01T), the corresponding magnetic energy is about 7×10^{28} ergs and the ohmic dissipation rate is of the order of 10^{10} W. Braginsky, a proponent of the "strong" toroidal field hypothesis, once calculated (Braginsky, 1965) a dissipation rate of 3.8×10^{12} W; he now suggests (Braginsky, 1976) a "more realistic" value 10 times smaller, or 4×10^{11} W.

A tenuous clue may be provided by Bukowinski's (1977) determination of the temperature gradient in the inner core, alluded to earlier (Chapter 3). By fitting "observed" properties of the inner core (density, seismic velocities) to a quantum-mechanical equation of state for iron, Bukowinski finds that the temperature at the center exceeds the temperature at the ICB by an amount ΔT, which varies from 234° to 350° according to the density model chosen; his preferred model PEM has $\Delta T = 234°$. In the steady state, a temperature gradient implies a heat source in the inner core that can hardly be other than the ohmic heating caused by electric currents diffusing from the outer into the inner core. The total rate Φ of ohmic heating in the inner core with radius r_i is then of the order of $4\pi r_i k \Delta T \cong 1 \times 10^{11}$ W for $k = 30$ W/m deg (this is an underestimate, since the temperature gradient at the ICB is likely to be steeper than the average gradient $\Delta T/r_i$ used here). If, then, we boldly assume the current density j to be uniform throughout the core, the rate of ohmic heating for the whole core comes out as $\Phi = \Phi_i (r_c/r_i)^3$, where r_c is the radius of the CMB; this gives $\Phi \cong 2.5 \times 10^{12}$ W, which seems high but could be reduced by a factor of 10 if the current density were, on the average, about 3 times greater in the inner core than in the outer core. It is also possible that the temperature gradient in the inner core also reflects the secular cooling postulated by the adherents of the gravitational dynamo. Perhaps all that can be said at the moment is that to balance ohmic dissipation, magnetic energy must be produced at a rate of 10^{10}–10^{11} W.

But how is magnetic energy created in a hydromagnetic dynamo? To see this, start from the expression for the current density in an ohmic conductor moving at velocity **u** relative to a magnetic field **B**:

$$\mathbf{j} = \mathbf{E} + (\mathbf{u} \times \mathbf{B}), \tag{4.7}$$

where **E** is the electric field such that curl $\mathbf{E} = -\partial \mathbf{B}/\partial t$. Dot **j** into the left side and the equivalent $(1/4\pi\mu)$ curl **B** into the right side. Using standard vector identities one obtains, after some algebra,

$$\frac{1}{\sigma}(\mathbf{j} \cdot \mathbf{j}) = -\frac{1}{4\pi\mu} \operatorname{div}(\mathbf{E} \times \mathbf{B}) + \frac{\partial}{\partial t}\left(\frac{B^2}{2}\right) - \mathbf{u} \cdot \mathbf{F}_L, \tag{4.8}$$

where $\mathbf{F}_L$ stands for the Lorentz force $(1/4\pi\mu)$ [(curl **B**) × **B**]. Multiply both sides by the element of volume dV and integrate over all space to infinity. The integral $\int \operatorname{div}(\mathbf{E} \times \mathbf{B}) \, dV = \int (\mathbf{E} \times \mathbf{B}) \cdot d\mathbf{S}$ goes to zero at infinity because both E and B decrease faster than r^2. Both **j** and **u** are zero outside the volume V of the conducting and flowing fluid. Thus we obtain

$$\frac{\partial E_m}{\partial t} = -\int_V \frac{j^2}{\sigma} dV - \int_V \mathbf{u} \cdot \mathbf{F}_L \, dV, \tag{4.9}$$

where $E_m = \int_V B^2/8\pi\mu \, dV$ is the total magnetic energy, inside and outside the fluid. The first integral on the right-hand side of (4.9) is of course the ohmic dissipation. The last term on the right, with its minus sign, is the rate at which the fluid does work against the Lorentz force.

Equation (4.9) shows that the rate of creation of magnetic energy equals the rate at which the fluid does mechanical work against the resistance offered by the Lorentz force, minus the rate of ohmic dissipation. In the steady state, $\partial E_m/\partial t = 0$; all the work done by the fluid is converted to heat by the electrical resistance of the conductor. Equation (4.9) also shows that when the fluid does no work (as when, for instance, the flow is parallel to $\mathbf{B}$, or more generally when $\mathbf{u}$ lies in the plane containing $\mathbf{B}$ and $\mathbf{j}$), the magnetic field decays at a rate determined by the ohmic dissipation.

Our problem now is to examine what conditions must pertain in the core to enable the fluid to do mechanical work against the Lorentz force at a rate at least equal to the ohmic dissipation (10^9–10^{11} W).

We consider first the case of a dynamo activated by thermal convection.

EFFICIENCY OF A STEADY-STATE THERMAL DYNAMO

Consider a convecting system receiving heat H at a rate $Q = dH/dt$. In the case of the core, Q could represent the rate of radioactive heat generation or the rate of release of latent heat by crystallization of the inner core. The problem is to determine the rate W at which the system can do mechanical work. The ratio $\eta = W/Q$ is called the efficiency of the system.

In classical thermodynamics, one usually considers a dissipationless system receiving heat at rate Q_1 from a source at temperature T_1, losing heat at rate Q_0 to a sink at temperature $T_0 < T_1$, and doing work *on the outside* at rate W. In the steady state, conservation of energy requires

$$W = Q_1 - Q_0. \tag{4.10}$$

The system is considered dissipationless, so that there is no production of entropy by irreversible processes (heat conduction, friction, etc.). The system receives entropy from the heat source at rate Q_1/T_1 and loses entropy to the heat sink at rate Q_0/T_0. In the steady state, entropy remains constant, so that

$$\frac{Q_1}{T_1} = \frac{Q_0}{T_0}. \tag{4.11}$$

Combining (4.10) and (4.11) gives for the efficiency

$$\eta = \frac{W}{Q_1} = 1 - \frac{Q_0}{Q_1} = \frac{T_1 - T_0}{T_1}, \tag{4.12}$$

which is necessarily smaller than 1 since $T_0 > 0$. Since any real system will be dissipative, (4.12) gives an upper bound to the efficiency.

Note that in (4.12) the work W must be done by the system on its surroundings. The MHD dynamo in the core is different, in the sense that in the steady state (constancy of B outside the core) no energy leaves the core other than the heat Q_0 transferred into the mantle ("the sink") at the temperature T_0 of the core-mantle boundary. As explained above, the work done by the fluid against Lorentz and viscous forces is converted back to heat *inside the core* by ohmic and viscous dissipation. Backus (1975) has shown that under these circumstances

$$W_m \leq \left(\frac{T_m}{T_0} - 1\right) Q, \tag{4.13}$$

where W_m is the sum of the rates of production of magnetic energy (which equals ohmic dissipation) and of viscous dissipation, T_0 is, as before, the temperature at the core-mantle boundary, T_m is the maximum temperature inside the core, and Q is the sum of the heat produced by distributed radioactive sources and of the heat that enters at the inner boundary (i.e., the inner-outer core boundary). Since there is in principle no reason why T_m cannot be greater than $2T_0$, there is, again in principle, no reason why the efficiency W_m/Q could not be greater than 1.

This somewhat paradoxical result may perhaps be understood by noting that since the ohmic and viscous heating occur within the convecting fluid, the heat generated by these dissipative processes could in principle also be used to power convection. Imagine for instance a system with a uniform distribution of radioactive sources in which dissipative heating is also uniform; an element of fluid cannot distinguish between the two sources, which therefore both contribute to the motion. The answer to the paradox is that the dissipative heating will not in general be uniform but will be distributed so as to oppose or cancel the temperature gradient required to drive the convection, i.e., to raise T_0 so that it approaches T_m. In a body with the low viscosity and high Reynolds number characteristic of the outer core, flow inside the core will be essentially inviscid, most of the viscous dissipation taking place in a thin boundary layer at the core-mantle interface. The same sort of thing will happen for the ohmic heating because of the high magnetic Reynolds number of the core. This may be seen for instance in Braginsky's (1976) "model Z" dynamo, in which electric currents flow mostly in a thin magnetic boundary layer where the magnetic field changes from its rather uniform axial character in the interior to its poloidal (mainly dipolar) form outside the core. In both cases, production of heat near the core-mantle boundary will tend to raise the temperature there and therefore reduce the negative temperature gradient that is needed for thermal convection. Clearly, if T_0 rises so that $T_0 \to T_m$, the efficiency given by (4.13) goes to zero. The author knows of no general theorem to prove that dissipation will occur so as to hinder the motion, but he strongly suspects that there must be one.

Metchnik et al. (1974) have considered the efficiency of convection in a layer of fluid heated from below. The flow is assumed to be isentropic so that the temperature T_2 at the top lies on the adiabat through T_1, the temperature at the bottom. Similarly, $T_2 + \Delta T_2$ lies on the adiabat through $T_1 + \Delta T_1$. They claim that the efficiency is

$$\eta = \frac{T_1 - T_2}{T_1}, \qquad (4.14)$$

as in (4.12). The result is erroneous. It is reached by assuming that since the flow is isentropic, the difference in entropy between an ascending column BD (Figure 4-2) and a descending column EA is the same at all heights; thus the difference in entropy between A and B is the same as between F and C. If this were true, it would also be the same as the entropy difference between D and E. The entropy difference between A and B arises from the heat received on the horizontal branch of the flow, which is $mc_P\Delta T_1$, where m is the mass of fluid and c_P its specific heat at constant pressure, and the entropy change ΔS_1 is $mc_P\Delta T_1/T_1$ (assuming that $\Delta T_1 \ll T_1$). Similarly, the entropy difference ΔS_2 is due to cooling along the segment DE and is $mc_P\Delta T_2/T_2$. Since the two are assumed to be equal, $\Delta T_1/T_1 = \Delta T_2/T_2$. But in the steady state no energy leaves the system other than the heat lost at the top, which must therefore equal the heat received at the bottom; hence $mc_P\Delta T_1 = mc_P\Delta T_2$ and $\Delta T_1 = \Delta T_2$. Then $T_1 = T_2$, which contradicts the assumption

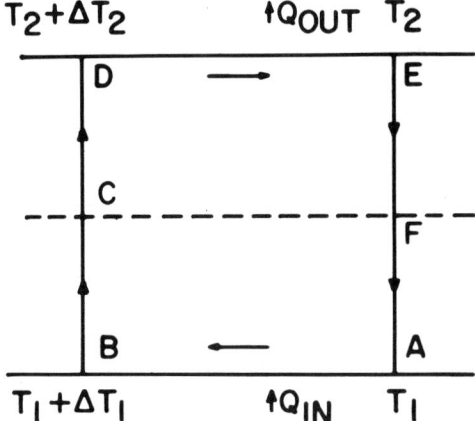

FIGURE 4-2 Temperatures in a rising limb (B to D) and a descending limb (E to A) of a convection cell in a layer of liquid heated from below. Metchnik et al. (1974) assume erroneously that the difference in specific entropy of the rising and falling fluids is the same at all levels (see text).

of an adiabatic gradient between A and E or between B and D; furthermore, if $T_1 = T_2$, the efficiency is zero by (4.14). The error in (4.14) stems from the assumption of isentropic flow. Since the system does no work on its surroundings, any work done (e.g., by buoyancy forces) inside the system must be dissipated within the system as viscous or ohmic heating, with a corresponding irreversible production of entropy. There is also the unavoidable irreversible production of entropy by thermal conduction at the rate $k(\Delta T/T)^2$ per unit volume, where k is the thermal conductivity. Both vertical and horizontal temperature differences (i.e., $T_2 - T_1$, ΔT_1, ΔT_2) contribute to this term.

AN IMPROVED ESTIMATE OF EFFICIENCY

Consider in particular the case of a core with distributed radioactive heat sources generating ϵ watts per unit volume, so that the total heat generation is $\int \epsilon \, dV$. In the steady state, the core must be losing heat to the mantle at the rate $Q_0 = \int \epsilon \, dV$, since no energy leaves the system under any form other than the heat flux through the core-mantle boundary, assumed to be held at the uniform temperature T_0.

The appropriate momentum equation, written in a coordinate system rotating at the uniform rate Ω, is

Dynamics of the Core

$$\rho \left[\frac{\partial \mathbf{u}}{\partial t} + (\mathbf{u} \cdot \nabla) \mathbf{u} \right] + 2\rho (\Omega \times \mathbf{u}) = -\nabla P + \rho \nabla \psi$$

$$+ \mathbf{F}_L + \mathrm{div}\, \tau, \tag{4.15}$$

where ρ is density, $\mathbf{u}$ is velocity, P is pressure, ψ is the sum of the gravitational and centrifugal potentials, $\mathbf{F}_L$ is the Lorentz force $\mathbf{j} \times \mathbf{B}$, and τ is the deviatoric stress tensor.

Forming the scalar product of both sides of (4.15) with $\mathbf{u}$ and integrating over the volume V of the fluid, one obtains after standard manipulations* (Hide, 1956), using (4.9),

$$\frac{\partial}{\partial t}(E_k + E_m) = \int_V P\, \mathrm{div}\, \mathbf{u}\, dV + \int_V \psi \frac{\partial \rho}{\partial t} dV$$

$$- \int_V \phi_m\, dV - \int_V \phi_v\, dV, \tag{4.16}$$

where $E_k = \tfrac{1}{2} \int_V \rho u^2/2\, dV$ is the kinetic energy of the fluid, $\phi_m = j^2/\sigma$ is the ohmic dissipation, and ϕ_v is the viscous dissipation. Let Φ be the total dissipation rate, and let $D = \int_V P\, \mathrm{div}\, \mathbf{u}\, dV$. In the steady state, when $\partial/\partial t = 0$, (4.16) reduces to the simple form

$$D = \phi = \int_V P\, \mathrm{div}\, \mathbf{u}\, dV, \tag{4.17}$$

which singles out D as the driving term for the dynamo. Since in the steady state $\int_V \phi_m\, dV = -\int_V \mathbf{u} \cdot \mathbf{F}_L\, dV$ by (4.9), the rate at which the fluid does work against the Lorentz force is measured by the same term D, and the efficiency η of the dynamo is, in the steady state,

$$\eta = D / \int \epsilon\, dV = D/Q_0 = \Phi/Q_0. \tag{4.18}$$

Equation (4.18) seems paradoxical since the efficiency increases with increasing rate of dissipation. The apparent paradox stems, as explained above, from the fact that since dissipation takes place within the system, it must be counted as an additional heat source.

---

*Repeated use is made of the identity $\mathrm{div}\, a\mathbf{B} = a\, \mathrm{div}\, \mathbf{B} + \mathbf{B} \cdot \nabla a$, where a is any scalar quantity and $\mathbf{B}$ is any vector.

Equation (4.17) also shows that a strictly divergenceless flow cannot maintain a dynamo. From the equation of continuity

$$\frac{\partial \rho}{\partial t} + \mathbf{u} \cdot \nabla \rho + \rho \text{ div } \mathbf{u} = 0, \tag{4.19}$$

it follows that, in the steady state, div $\mathbf{u} = 0$ implies $\nabla \rho = 0$, which cannot be true in a real fluid subjected to pressure and temperature gradients. It is nevertheless common practice to write div $\mathbf{u} = 0$, as if the density were uniform, and to retain the density variation $\delta \rho$ only in the buoyancy term $\rho \nabla \psi$ of the momentum equation. This, the Boussinesq approximation, is known to be valid provided that $\delta \rho / \rho \ll 1$, or if the thickness d of the convecting layer of fluid is much smaller than the temperature scale height $h_T = c_P / g\alpha$ (Hewitt et al., 1975). If we take for the earth's core $c_P \cong 0.6$ J/g deg, $\alpha \cong 10^{-5}$/deg, and $g \cong 10^3$ cm/s^2, we find $h_T = 6 \times 10^8$ cm, and the ratio $d/h_T \cong \frac{1}{3}$, which is not small. Recall also that density varies by about 2 g/cm^3 between the top and bottom of the outer core with a mean density of about 10 g/cm^3, so that $\delta \rho / \rho \cong 0.2$, which is not small either. Thus the applicability of the Boussinesq approximation for the earth's core is questionable. It seems nevertheless that div $\mathbf{u}$ will be small if not neglible. Equation (4.18) then predicts that the efficiency of the dynamo will also be small, if not exactly zero.

The important term D can be put under a variety of forms. We now spell out some of them.

Since P div $\mathbf{u}$ = div $(P\mathbf{u}) - \mathbf{u} \cdot \nabla P$ and the normal component of $\mathbf{u}$ vanishes on the core boundary,

$$D = \int_V P \text{ div } \mathbf{u} \, dV = -\int_V \mathbf{u} \cdot \nabla P \, dV, \tag{4.20}$$

which shows that flow along isobaric surfaces contributes nothing to D or to Φ. Alternatively, since in the steady state div $\mathbf{u} = -(\mathbf{u} \cdot \nabla \rho)/\rho$, we also have

$$D = -\int_V \frac{P}{\rho} \mathbf{u} \cdot \nabla \rho \, dV, \tag{4.21}$$

which shows that flow along surfaces of equal density contributes nothing to D.

Since $\nabla \rho / \rho = -\alpha \nabla T + \nabla P / K_T$, where K_T is the isothermal bulk modulus, (4.21) can be further transformed to

$$D = \int_V \alpha P \mathbf{u} \cdot \nabla T \, dV - \int_V \frac{P}{K_T} \mathbf{u} \cdot \nabla P \, dV. \qquad (4.22)$$

Still other forms can be obtained for D from the principle of local equilibrium (Glansdorff and Prigogine, 1971, p. 14), according to which the local values of the thermodynamic variables T, P, and $\underline{v} = 1/\rho$ can be defined as if the system were in thermodynamic equilibrium:

$$T = \left(\frac{\partial e}{\partial s}\right)_{\underline{v}}, \qquad -P = \left(\frac{\partial e}{\partial \underline{v}}\right)_s, \qquad \left(\frac{\partial s}{\partial \underline{v}}\right)_e = P/T,$$

where e is the specific internal energy and s is the specific entropy. Then

$$T \, ds = de + P \, dv \qquad (4.23)$$

and since $dv/dt = -(1/\rho^2) \, d\rho/dt = (1/\rho) \, \text{div } \mathbf{u}$, (4.23) can be written as

$$P \, \text{div } \mathbf{u} = T\rho \frac{ds}{dt} - \rho \frac{de}{dt}. \qquad (4.24)$$

Making use of (4.19) it can also be shown that (Glansdorff and Prigogine, 1971, p. 4)

$$\rho \frac{ds}{dt} = \rho \left(\frac{\partial s}{\partial t} + \mathbf{u} \cdot \nabla s\right) = \frac{\partial}{\partial t}(\rho s) + \text{div}(\rho s \mathbf{u}),$$

$$\rho \frac{de}{dt} = \rho \left(\frac{\partial e}{\partial t} + \mathbf{u} \cdot \nabla e\right) = \frac{\partial}{\partial t}(\rho e) + \text{div}(\rho e \mathbf{u}). \qquad (4.25)$$

Thus from (4.24) and (4.25)

$$D = \int_V T\rho \frac{ds}{dt} dV - \int_V \rho \frac{de}{dt} dV$$

$$= \int_V T \, \text{div}(\rho s \mathbf{u}) \, dV$$

$$- \int_V \text{div}(\rho e \mathbf{u}) \, dV.$$

The last integral on the right can be transformed to a surface integral that vanishes because the normal component of **u** is zero on the boundary. The other integral can also be transformed, noting that $T \operatorname{div}(\rho s \mathbf{u}) = \operatorname{div}(T\rho s \mathbf{u}) - \rho s \mathbf{u} \cdot \nabla T$, to yield

$$D = -\int_V \rho s \mathbf{u} \cdot \nabla T \, dV \tag{4.26}$$

or by the first of the equations (4.25)

$$D = \int_V T\rho \mathbf{u} \cdot \nabla s \, dV , \tag{4.27}$$

which shows that isothermal or isentropic flows do not contribute to D.

Finally, if we replace ∇s in (4.27) by the equivalent $(c_P/T)\nabla T - (\alpha/\rho)\nabla P$, we get

$$D = \int_V \rho c_P \mathbf{u} \cdot \nabla T \, dV - \int_V \alpha T \mathbf{u} \cdot \nabla P \, dV . \tag{4.28}$$

If c_P is constant, the first integral vanishes because $\rho \mathbf{u} \cdot \nabla T = \operatorname{div}(\rho T \mathbf{u}) - T \operatorname{div}(\rho \mathbf{u})$, $\operatorname{div}(\rho \mathbf{u}) = 0$ in the steady state, and $\int_V \operatorname{div}(\rho T \mathbf{u}) \, dV = 0$ because of the boundary condition on **u**. Thus

$$D = -\int_V \alpha T \mathbf{u} \cdot \nabla P \, dV , \tag{4.29}$$

which is the form used by Hewitt et al. (1975) to estimate the efficiency of the core dynamo. They proceed as follows: First split P into a hydrostatic term P_0 such that $\nabla P_0 = \rho g$ and a dynamic term P_1, with $P = P_0 + P_1$. If the Reynolds number of the flow is small, $\nabla P_1 \ll \nabla P_0$ and can be neglected. Let w be the radial velocity. Then

$$D = \Phi = \int_V \rho g \alpha T w \, dV = \int_0^a \frac{g\alpha}{c_P} F(r) \, dr , \tag{4.30}$$

where $F(r)$ is the convective heat flux $4\pi r^2 \rho c_P \langle Tw \rangle$ at radius r (the brackets denote averages over spherical surfaces) and a is the outer radius of the core. Suppose further that $g(r) = rg_0/a$, where g_0 is the value of g on $r = a$. If heat is generated uniformly within the sphere and is carried by convection only,

$$F(r) = \left(\frac{r}{a}\right)^3 Q_0 , \tag{4.31}$$

where Q_0 is the rate at which heat is supplied at the boundary. Substituting in (4.30) gives

$$\Phi = \frac{1}{5}\left(\frac{g_0 \alpha a}{c_P}\right) Q_0$$

and the efficiency

$$\eta = \frac{\Phi}{Q_0} = \frac{1}{5}\frac{a}{h_T}. \qquad (4.32)$$

Since $a \cong 3.5 \times 10^8$ cm and $h_T \cong 6 \times 10^8$ cm, $\eta \cong 0.12$.

This value of η is likely to be an overestimate, since the calculation leading to it ignores the conductive contribution to heat transport. We may note in the first place that since $\int_V \mathbf{u} \cdot \nabla P_0 dV = 0$ in the steady state, the integral (4.29) would be zero, and η would be zero, if α and T were constant; but if T is not constant, ∇T is not zero everywhere, and heat cannot be carried by convection only. Clearly, Equation (4.31) is grossly wrong at $r = a$, where it requires $F(r) = Q_0$; but on $r = a$ the convective heat flux $F(r)$ is necessarily zero since w vanishes there. It is precisely in the thermal boundary layer near $r = a$ that ∇T is likely to be largest and the entropy production by conduction, which is proportional to $(\nabla T)^2$ is likely to be greatest. As we shall see in the next section, this conductive contribution to entropy production decreases the efficiency below the value it might have in the absence of conduction and makes it impossible to determine the efficiency in the absence of detailed information on the temperature distribution. This can also be seen from the entropy balance equation, to which we now turn.

THE ENTROPY BALANCE EQUATION

Suppose that radioactivity is the only source of heat in the core. In the steady state, the entropy of the core must remain constant, even though it is losing entropy at the rate Q_0/T_0, where T_0 is the temperature in the CMB and $Q_0 = \int_V \epsilon \, dV$, ϵ being the rate of radiogenic heat production per unit volume. The entropy loss must be balanced by irreversible entropy production within the core. There are three internal sources of entropy, namely radiogenic heat production, dissipation (ohmic and viscous), and heat conduction. Thus entropy balance requires that

$$\frac{Q_0}{T_0} = \frac{1}{T_0}\int_V \epsilon \, dV$$

$$= \int_V \frac{\epsilon}{T} dV + \int_V \frac{\phi_m}{T} dV$$

$$+ \int_V \frac{\phi_v}{T} dV + \int_V k \left(\frac{\nabla T}{T}\right)^2 dV , \qquad (4.33)$$

where k is, as before, the thermal conductivity, ϕ_m is the ohmic dissipation rate, and ϕ_v is the viscous dissipation rate. Thus

$$\int_V \frac{\phi_m}{T} dV =$$

$$\int_V \epsilon \left(\frac{1}{T_0} - \frac{1}{T}\right) dV - \int_V k \left(\frac{\nabla T}{T}\right)^2 dV$$

$$- \int_V \frac{\phi_v}{T} dV . \qquad (4.34)$$

Clearly, a maximum value for ϕ_m/T is obtained if the fluid is inviscid and $\phi_v = 0$, which we shall assume to be the case. Let T_m be the maximum temperature within the body, so that $T_0 \leq T \leq T_m$. Then, since ϕ_m is necessarily positive everywhere,

$$\frac{1}{T_m} \int_V \phi_m \, dV \leq \int_V \frac{\phi_m}{T} dV \leq \frac{1}{T_0} \int_V \phi_m \, dV$$

or

$$T_0 \int_V \frac{\phi_m}{T} dV \leq \int_V \phi_m \, dV \leq T_m \int_V \frac{\phi_m}{T} dV . \qquad (4.35)$$

Substituting the value of $\int_V (\phi_m/T) \, dV$ from (4.34) in (4.35) yields

$$\int_V \epsilon \, dV = T_0 \int_V \frac{\epsilon}{T} dV - T_0 \int_V k \left(\frac{\nabla T}{T}\right)^2 dV \leq \int_V \phi_m \, dV$$

$$\leq \frac{T_m}{T_0} \left[\int_V \epsilon \, dV - T_0 \int_V \frac{\epsilon}{T} dV - T_0 \int_V k \left(\frac{\nabla T}{T}\right)^2 dV \right] .$$

$$(4.36)$$

The efficiency η is defined as before,

$$\eta = \int \phi_m \, dV / \int \epsilon \, dV = \Phi/Q_0 \, .$$

Suppose that ϵ is uniform, so that $\int_V \epsilon \, dV = \epsilon V$. Dividing (4.36) throughout by ϵV yields

$$1 - \lambda \leq \eta \leq \frac{T_m}{T_0}(1 - \lambda), \tag{4.37}$$

where

$$\lambda = \frac{T_0}{V}\left[\int_V \frac{dV}{T} + \frac{1}{\epsilon}\int_V k\left(\frac{\nabla T}{T}\right)^2 dV\right]. \tag{4.38}$$

Equations (4.37) and (4.38) clearly show the role of the entropy production by conduction of heat, $\dot{S}_c = k \int_V (\nabla T/T)^2 \, dV$. Define an average temperature T_a as

$$\frac{1}{T_a} = \frac{1}{V}\int_V \frac{dV}{T}. \tag{4.39}$$

Then (4.38) can be written as

$$\lambda = \frac{T_0}{T_a} + \frac{\dot{S}_c}{\dot{S}}, \tag{4.40}$$

where $\dot{S}$ is the total rate of entropy production $Q_0/T_0 = \epsilon V/T_0$. Substituting (4.40) into (4.37) leads to an upper bound on η

$$\eta \leq \frac{T_m}{T_0}\left(1 - \frac{T_0}{T_a} - \frac{\dot{S}_c}{\dot{S}}\right) = T_m \frac{T_a - T_0}{T_a T_0} - T_m \frac{\dot{S}_c}{Q_0}, \tag{4.41}$$

which shows that the upper bound to η decreases as the conduction entropy production increases. This upper bound is, incidentally, always below Backus' (1975) upper bound $(T_m - T_0)/T_0$, since $\dot{S}_c \geq 0$ and $T_a \leq T_m$.

Using the identity

$$\text{div}\left(\frac{\nabla T}{T}\right) = \frac{1}{T}\nabla^2 T - \frac{1}{T^2}(\nabla T)^2$$

and noting that

$$\int_V -k \,\text{div}\left(\frac{\nabla T}{T}\right) dV = \int_V -k\,\frac{\nabla T \cdot \mathbf{dS}}{T_0} = \frac{Q_0}{T_0} = \frac{\epsilon V}{T_0},$$

(4.38) can be rewritten as

$$\lambda = 1 + J,$$

where J is defined as

$$J = \frac{T_0}{\epsilon V} \int_V \frac{(\epsilon + k\nabla^2 T)}{T}\, dV = \frac{T_0}{T_a} + \frac{T_0}{Q_0}\int_V \frac{k\nabla^2 T}{T}\, dV. \quad (4.42)$$

Then (4.37) becomes

$$-J \leq \eta \leq -\frac{T_m}{T_0} J \quad (4.43)$$

and since η is necessarily positive or zero (since ϕ and ϵ are positive or zero everywhere), J must be negative. If $J = 0$, $\eta = 0$; this happens when

$$1 + \frac{k}{\epsilon}\nabla^2 T = 0,$$

which describes the steady-state conductive temperature distribution in the absence of convection. Note that $\int_V (\epsilon + k\nabla^2 T)\, dV = 0$ if k is constant. Indeed,

$$\int_V k\nabla^2 T\, dV = \int_V k\,\text{div}\,(\nabla T)\, dV = \int_S k\nabla T \cdot \mathbf{dS}$$

$$= -\int_S \mathbf{q}_0 \cdot \mathbf{dS} = -Q_0 = -\int_V \epsilon\, dV,$$

where $\mathbf{q}_0$ is the heat flux at the surface. Thus efficiency is proportional to $\int_V [(\epsilon + k\nabla^2 T)/T]\, dV$, with $\int_V (\epsilon + k\nabla^2 T)\, dV = 0$. Clearly, J and limits on the efficiency cannot be calculated in the absence of precise knowledge of the temperature distribution and of $\nabla^2 T$, which in turn requires that the dynamo problem first be solved. No exact solutions are available yet.

If we try to estimate η by choosing a "likely" yet unproven temperature distribution, we recall that Φ and η are zero if (1) the temperature is a solution of the conduction equation, or if (2) the temperature profile is adiabatic so that $\nabla s = 0$ everywhere (see Equation (4.27)). Φ is also zero if the temperature is a function of r only. Indeed, in a steady state, constancy of mass inside a spherical surface S of radius r requires that the mass flux across the surface be zero, i.e.,

$$\int_S \rho \mathbf{u} \cdot d\mathbf{S} = 0 \ .$$

But if T is constant on S, so is ρ, so that we must have

$$\int_S \mathbf{u} \cdot d\mathbf{S} = \int_{V(r)} \text{div } \mathbf{u} \, dV = 0 \ ,$$

where the volume integral is over the volume $V(r)$ inside S. Since this must be true for any r, it follows that div $\mathbf{u} = 0$ everywhere, and $D = \Phi = 0$ (Equation (4.17)). The production of magnetic energy thus depends critically on the horizontal temperature gradients, without which there would obviously be no convection.

The best we can do at the moment is to assume $\nabla^2 T = -T/l^2$ where l is a length characteristic of the scale of the temperature fluctuations. Then

$$J = \frac{T_0}{T_a} - \left(\frac{T_0}{Q_0}\right)\left(\frac{kV}{l^2}\right) .$$

Substituting this in (4.43) and solving for Q_0 gives

$$T_a \frac{kV}{l^2} - \Phi \frac{T_a}{T_0} \leq Q_0 \leq T_a \frac{kV}{l^2} - \Phi \frac{T_a}{T_m} , \qquad (4.44)$$

for which the heat generation in the core, Q_0, could be estimated for any value of Φ if a reasonable choice could be made for l. Take $k = 30$ W/m deg (Stacey, 1977), $T_0 = 4500°K$, $T_a = 5000°K$, $T_m = 6000°K$, and $\Phi = 4 \times 10^{11}$ W. Then, for $l = 10^6$ m,

$$2.51 \times 10^{13} \leq Q_0 \leq 2.52 \times 10^{13} ,$$

implying an efficiency of less than 2 percent; but no real significance can be attached to this number, which depends on guessing at the value of l. However, since l is unlikely to be much greater than 3×10^6 m (roughly the radius of the outer core), Q_0 is not likely to be much less than 2.6×10^{12} W. For comparison, we recall that 0.1 percent by weight of potassium in the outer core produces about 6.7×10^{12} W ($\epsilon \cong 4 \times 10^{-8}$ W/m³).

To recapitulate: it is not possible to calculate the efficiency of the dynamo (i.e., the ratio of ohmic dissipation Φ to total heat output $Q_0 = \epsilon V$) without a detailed knowledge of the temperature distribution in the convecting fluid. An upper limit of about 0.1 was calculated by Hewitt *et al.* (1975) by ignoring conduction and assuming that heat is carried by convection only. However, the effect of conduction, namely the entropy production associated with it, is not small if temperatures in the core are anywhere near the values previously determined (about 4500°K at the outer core-mantle boundary, about 6000°K at the inner core boundary), so that the actual efficiency may be an order of magnitude smaller than the Hewitt *et al.* (1975) value. An approximate evaluation of the actual efficiency is given by Equation (4.44), from which it would be possible to calculate the heat output Q_0, given the ohmic dissipation Φ and the length scale l of the horizontal temperature fluctuations defined by $\nabla^2 T = -T/l^2$. For $l = 1 \times 10^6$–3×10^6 m, Q_0 is in the range 2.8×10^{12}–2.5×10^{13} W.

THE GRAVITATIONAL DYNAMO

Braginsky's early dynamo models required a very large toroidal field and had the correspondingly large dissipation rate of 3.8×10^{12} W. Noting that the efficiency of a thermal dynamo would necessarily be low, he concluded that convection in the core could not be thermal, as it would require that more heat be generated in the core per unit time than escapes at the earth's surface. He suggested instead (Braginsky, 1963) that convection is powered gravitationally, either by the floating upward of the excess of the light component (silicon in those days) released as the inner core crystallizes, or by heavy material falling from the mantle into a growing core. Verhoogen (1961) had already suggested that the release of latent heat of crystallization could contribute significantly to maintenance of convection. The gravitational dynamo has been reconsidered lately by Gubbins (1977) and by Loper (1978). The problem is difficult.

The general principle of the gravitational dynamo is as follows. Cooling of the core must cause an outward displacement of the inner core boundary; the inner core grows by crystallization of iron. Since the density of the inner core is greater than the density of the outer core, growth of the inner core entails a release of gravitational energy. As a result of crystallization of iron, a layer of liquid depleted in iron and therefore enriched in sulfur forms at the ICB. This layer is assumed to be less dense than the rest of the outer core. It rises by buoyancy to produce the convective motion that generates electric currents.

GRAVITATIONAL ENERGY

Loper (1978) has calculated the amount of gravitational energy released by comparing the gravitational energy of the earth as it is today to the gravitational energy of the hotter earth just prior to beginning of crystallization. He constructs a model of a compressible earth in which the mass of the inner core is an independent variable, using an approximate equation of state based on present properties of the earth and making no allowance for the higher temperatures prevailing before crystallization began. He calculates a total release to date of approximately 2.5×10^{29} J. If the inner core began to form 4.5 billion years ago, the average rate of release is 1.76×10^{12} W; if the age of the inner core is only 3×10^9 yr, the rate is 2.64×10^{12} W. These numbers are very uncertain, since the gravitational energy (2.5×10^{29} J) is the difference between two large numbers, both of the order of 10^{32} J, and both uncertain by at least 10 percent, or perhaps even 50 percent. Central condensation of matter in the core causes g to rise, so that the pressure in the core also rises, and so does the temperature. Part of the gravitational energy thus goes into heat of adiabatic compression. Loper, comparing results for compressible and incompressible earths, estimates that not more than 27 percent of the power goes into internal heating and elastic compression; this, Loper says, leaves 1.28×10^{12} W to drive the dynamo, which seems ample. This, however, neglects all dissipative processes, other than ohmic heating, by which the gravitational energy could be converted to heat. That such dissipative processes do exist is beyond doubt, as we can see by asking ourselves where the gravitational energy would go if the outer core consisted of pure iron, so that no buoyant layer could form, or if the electrical resistivity of the outer core happened to be so large that electrical currents could not flow. Where, for instance, did the much larger amount of gravitational energy released by separation of mantle and core (Chapter 2) go?

A more precise evaluation of the gravitational input to the dynamo may be obtained by returning to the momentum equation (4.15), forming its dot product with velocity **u** and integrating over the volume of the core. Assuming the magnetic and kinetic energies to be constant, we obtain as before for the driving term D,

$$D = \Phi = -\int_V \mathbf{u} \cdot \nabla P \, dV + \int_V \rho \mathbf{u} \cdot \nabla \psi \, dV , \quad (4.45)$$

where Φ stands for the total dissipation, ohmic plus viscous plus whatever frictional dissipation may occur. This is Gubbins' Equation (7) (Gubbins, 1977). After some simple transformations, we get

$$D = \int_V P \operatorname{div} \mathbf{u} \, dV - \int_S P \mathbf{u} \cdot d\mathbf{S} + \int_S \rho \psi \mathbf{u} \cdot d\mathbf{S}$$
$$+ \int_V \psi \frac{\partial \rho}{\partial t} dV , \quad (4.46)$$

where the surface integrals are over the CMB.

The last three terms on the right of (4.46) were dropped in (4.17) because of the assumption $\partial \rho / \partial t = 0$, which is no longer valid, and because the normal component of velocity on the CMB was assumed to be zero. This, however, would no longer be the case if, as a result of crystallization of iron, the volume of the whole core were to change, thereby moving the CMB outward if the core expands, or inward if the core contracts (see below).

A slight simplification is obtained in (4.46) if, following Gubbins (1977), we replace ψ by $\psi_s + \psi_r$, where ψ_s is the value of the potential on the CMB, which is taken to be an equipotential surface. Then

$$\int_V \psi_s \frac{\partial \rho}{\partial t} dV = \psi_s \int_V \frac{\partial \rho}{\partial t} dV = -\psi_s \int_V \operatorname{div}(\rho \mathbf{u}) \, dV$$
$$= -\psi_s \int_S \rho \mathbf{u} \cdot d\mathbf{S} = -\int_S \rho \psi \mathbf{u} \cdot d\mathbf{S} \quad (4.47)$$

so that, finally,

$$D = \Phi = \int_V P \operatorname{div} \mathbf{u} \, dV - \int_S P \mathbf{u} \cdot d\mathbf{S} + \int_V \psi_r \frac{\partial \rho}{\partial t} dV . \quad (4.48)$$

The first integral on the right of (4.48) is the same as before, except that variations in density are now caused by compositional differences rather than by changes in temperature. The two other terms specifically represent the gravitational contribution to the dynamo. If the whole core contracts while it cools and crystallizes, so that $\mathbf{u}$ is directed inward, the pressure term represents work done by the mantle falling in, so to speak, on the shrinking core. If, on the other hand, the core expands (see below), this same term represents work that must be done to lift the mantle and is a negative contribution to the dynamo.

The last term on the right of (4.48) is estimated by Gubbins to be 1.7×10^{11} W. This evaluation is based on an assumed rate of crystallization of 25 m^3/s; at that rate it would take the inner core 10^{10} yr to grow to its present size, so that Gubbins' value may be an underestimate.

To estimate the second term on the right-hand side of (4.48) we must know the rate u at which the core boundary moves. This requires some consideration of volumetric relations.

VOLUMETRIC RELATIONS

It is customary in calculations pertaining to the core to suppose that the iron-sulfur melt behaves as a perfect binary solution, in which, by definition, the two liquid end members (e.g., iron and FeS) mix in all proportions without change in volume. Unfortunately, the Fe-FeS system is not a perfect situation, at least at low pressure. This is shown, for instance, by the fact that below 50 kbar, pressure has very little effect on the eutectic temperature, implying that the volume of the eutectic liquid is very nearly equal to the volume of a mixture of solids in the eutectic proportion. Since both pure iron and pure FeS melt with an increase in volume, contraction of the liquid must occur when the two pure liquids are mixed. Whether this effect persists at high pressure is not known. It is known that at pressures greater than about 55 kbar the eutectic temperature begins to rise with increasing pressure, but this effect may be due to a phase change in solid FeS.

Consider now a melt containing n_1 mol of iron (molecular mass M_1) and n_2 mol of FeS (molecular mass M_2). The volume V_0 of the melt is

$$V_0 = n_1 \bar{v}_1 + n_2 \bar{v}_2, \tag{4.49}$$

and its density ρ_0 is

$$\rho_0 = \frac{n_1 M_1 + n_2 M_2}{n_1 \bar{v}_1 + n_2 \bar{v}_2}, \tag{4.50}$$

where $\bar{v}_1$ and $\bar{v}_2$ are, respectively, the partial molar volumes of iron and FeS in the melt.

Suppose now that n_1 is changed by an amount δn_1, corresponding for instance to crystallization of δn_1 mol of iron that separate from the melt. From (4.50),

$$\frac{\partial \rho_0}{\partial n_1} = \frac{n_2}{V_0^2} (\bar{v}_2 M_1 - \bar{v}_1 M_2), \tag{4.51}$$

assuming that δn_1 is sufficiently small that $\bar{v}_1$ and $\bar{v}_2$ do not change appreciably. The formation of a buoyant layer lighter than the remaining liquid requires that ρ_0 decrease when iron is taken out of the melt, or $\partial \rho_0 / \partial n_1 > 0$. This, by (4.51), requires

$$\frac{\bar{v}_2}{\bar{v}_1} > \frac{M_2}{M_1} \cong \frac{90}{58} = 1.55. \tag{4.52}$$

If δn_1 mol of iron crystallize out of the liquid, the volume V_s of the solid so formed is $V_s = v_s \, \delta n_1$, v_s being the molar volume of the solid iron. The volume V_l of the remaining liquid is

$$V_l = (n_1 - \delta n_1) \bar{v}_1 + n_2 \bar{v}_2,$$

and the total volume V of solid and liquid is

$$V = V_s + V_l = V_0 + (v_s - \bar{v}_1) \delta n_1. \tag{4.53}$$

The volume of the whole core increases if $v_s > \bar{v}_1$. Contraction occurs if $v_s < \bar{v}_1$.

To see what could happen in the core, it is instructive to plot molar volume against composition in a binary system (Figure 4–3). On the ordinate axis $N_2 = 0$ (pure iron), we plot at A the molar volume v_1^0 of pure liquid iron under the pressure and temperature conditions considered; similarly at B we plot the molar volume v_2^0 of pure liquid FeS. The dashed line AB represents the molar volume of a perfect solution in which neither contraction nor expansion occurs on mixing. If contraction occurs, as it does in the Fe-FeS system at low pressure, the molar volume of the melt is represented by the curve APB, the exact shape of which is not known. It is a general property of such molar diagrams that the tangent at any point P to the curve cuts the two ordinate axes at points representing the two partial molar volumes $\bar{v}_1$ and $\bar{v}_2$, respectively. It is clear from the graph that if departures from perfect behavior are serious, $\bar{v}_2$ could be negative at low FeS concentration. It is also clear that at higher FeS concentrations—say, near the minimum of the APB curve—$\bar{v}_1$ could be very much smaller than v_1^0; and since at core pressures pure iron melts with relatively small change in volume (Leppaluoto, 1973), $\bar{v}_1$ could conceivably be smaller than v_s.

This, however, is unlikely. The configuration of the core (solid inner

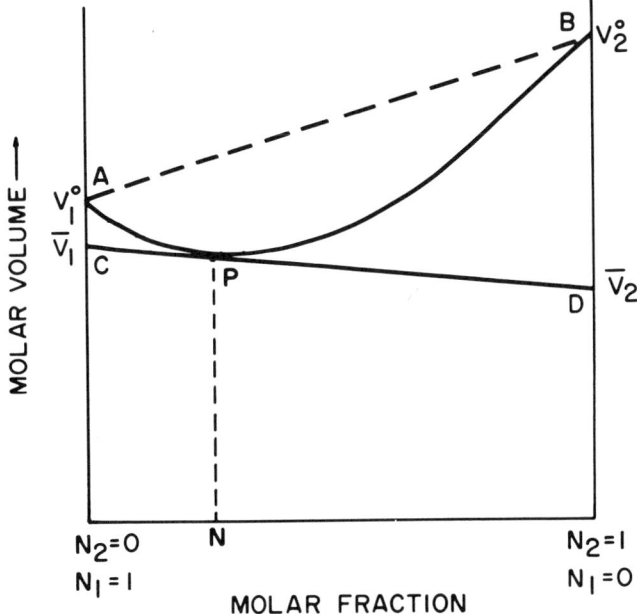

FIGURE 4–3 Molar volume diagram for a binary solution with negative volume of mixing. The tangent at P to the curve APB intersects the two ordinate axes at points representing, respectively, the partial molar volumes of the two components in a solution with molar composition N.

core inside a liquid outer core) requires that the liquidus temperature (i.e., the temperature T_m at which solid pure iron is in equilibrium with a Fe-FeS melt) should increase with increasing pressure. Now, at constant composition

$$\left(\frac{\partial T_m}{\partial P}\right)_N = \frac{\bar{v}_1 - v_s}{\bar{s}_1 - s_s},$$

where $\bar{s}_1$ and s_s are, respectively, the partial molar entropy of iron in the melt and the molar entropy of solid iron. If $\bar{v}_1 - v_s < 0$, $(\partial T_m/\partial P)_N$ can be positive only if $\bar{s}_1 - s_s > 0$, which implies that crystallization absorbs heat; if so, crystallization could not be induced by cooling. It is much more likely that $\bar{v}_1 > v_s$, the difference $\bar{v}_1 - v_s$ being of the same order as the difference $\Delta v = v_1{}^0 - v_s$, between the molar volumes of pure liquid and pure solid iron; Leppaluoto (1972b) estimates $\Delta v = 0.055$ cm³/mol at the pressure of the inner core boundary.

The volume change sustained by the whole core since crystallization of the inner core began some time Δt ago is then

$$\Delta V = -(\bar{v_1} - v_s)\,\Delta n,$$

where $\Delta n = M_i/M_1$, M_i being the mass of the inner core, approximately 9.8×10^{25} g. Thus $\Delta V \cong -9.3 \times 10^{22}$ cm^3.

A much larger contraction results from cooling, if our estimate of 250° for the average cooling ΔT of the core since crystallization began is correct (see below). If the average coefficient of thermal expansion α is taken to be 1×10^{-5}/deg (Stacey, 1977), the contraction amounts to $\Delta V = \alpha V\,\Delta T = -4.3 \times 10^{23}$ cm^3. For $\Delta t = 4 \times 10^9$ yr $= 1.26 \times 10^{17}$ s, the velocity of the boundary $u = da/dt = -\alpha a\,\Delta T/3\,\Delta t = -2.3 \times 10^{-12}$ cm/s, where a is the radius of the core; this corresponds to an inward displacement of the core boundary of 2.9 km over 4 billion years.

A further contraction results from the increase in pressure caused by the central condensation of matter. Loper (1978) estimates that the pressure at $r = 0$ may have risen by some 0.23 Mbar since the inner core began to form, enough to raise locally the density by more than 1 percent. Thus our value of $\Delta V = -4.3 \times 10^{23}$ cm^3 may be seriously underestimated.

A simple mechanism for converting gravitational energy into kinetic energy is by formation on the ICB of a layer of fluid lighter than the rest of the outer core liquid. This, by (4.52), requires

$$\frac{\bar{v_2}}{\bar{v_1}} > \frac{M_2}{M_1}. \tag{4.54}$$

This condition is likely to be satisfied on the whole, since the density of the outer core is assumed to be less than that of pure liquid iron precisely because of the addition of sulfur. But recall from Figure 4–3 that $\bar{v}_1$ and $\bar{v}_2$ are both likely to be sensitive to composition; small or even negative values of $\bar{v}_2$ are not excluded at low FeS contents. Since the shape of the curve APB in Figure 4–3 is not even approximately known, the possibility cannot be excluded *a priori* that (4.54) not be satisfied for certain compositions, including the actual composition of the core. The formation of a buoyant layer is therefore not certain, even though it appears likely. Formation of a buoyant layer also requires that diffusion of sulfur (or FeS) be sufficiently slow to prevent equalization of composition before the buoyant layer has had time to rise. Finally, we must choose to ignore the possibility, pointed out earlier by

Verhoogen (1973), that two immiscible liquids with different sulfur content might form. Clearly, a lot of experimental work on the Fe-FeS system at high pressure is needed.

We return to the evaluation of terms in Equation (4.48). The pressure on the CMB being about 1.4 Mbar, the surface pressure integral amounts to 4.9×10^{11} W, or more if we have underestimated ΔV and da/dt. This term does not contribute directly to the dynamo; being a measure of the work done in the core by its surroundings (i.e., the mantle), it goes into internal energy and heat, slowing down the rate of cooling of the core, so that it must be retained when we later consider the rate at which the core is losing heat.

There remains to evaluate the first term on the right-hand side of (4.48). Here div $\mathbf{u} = -(1/\rho) \, \partial\rho/\partial t - (1/\rho)\mathbf{u} \cdot \nabla\rho$ is a function of the density variations induced by crystallization of the core, by formation of a buoyant layer, and by the temperature gradient that must necessarily exist since the core is assumed to be cooling. There is no simple way of evaluating the integral.

All that can be said at the moment is that the only gravitational contribution to the dynamo that can be approximately evaluated is the term $\int_V \psi_r (\partial\rho/\partial t) \, dV$, which is probably larger than Gubbins' estimate of it (1.7×10^{11} W) and presumably sufficient to maintain the dynamo if, as Gubbins claims, gravitational energy released by rearrangement of matter in the core is completely converted to magnetic dissipation. That claim, however, can hardly be sustained at the moment. Clearly, the same release of gravitational energy by rearrangement of matter could occur in a nonconducting fluid in which no current can flow and no ohmic dissipation is permitted and in which other nonohmic dissipative processes would necessarily occur; these might also be operative in the earth's core. Evaluation of efficiency would, however, be even more difficult than for the thermal dynamo, because of chemical diffusion. Just as irreversible entropy production by conduction of heat turned out to be an important factor in the thermal dynamo, irreversible entropy production by chemical diffusion in a fluid of varying composition could well limit the efficiency of the chemical dynamo.

HEAT OUTPUT OF THE CORE

We now attempt to estimate the rate at which the core must be losing heat for the gravitational dynamo to operate. The heat output will consist of (1) the released gravitational energy transformed into heat by ohmic heating and other forms of dissipation, and (2) the heat released by cooling of the core and crystallization of the inner core. The first source we have found to be greater than 6.6×10^{11} W; we now proceed to evaluate the second.

We start at the moment when the temperature at the center of the earth has cooled down to the solidus temperature appropriate to the pressure and composition of the core. Suppose the core contains 10 percent sulfur by weight (= 28.1 percent FeS), the molar fraction x_1 of iron being 0.8. To make the calculation at all feasible, we must now assume that the melt behaves as a perfect solution. The solidus temperature T_m at molar fraction x_1 is

$$T_m = \frac{1}{R} \frac{\Delta h^0}{\left(\dfrac{\Delta h^0}{RT_1^0} - \ln x_1\right)}, \qquad (4.55)$$

where Δh^0 is the latent heat of pure component 1 and T_1^0 is its melting point. For pure iron at $P = 3.3$ Mbar, Leppaluoto (1972) estimates $T_1^0 = 7400°$K, $\Delta h^0 = T_1^0 \Delta s_1^0 = 3560$ cal/mol. For $x_1 = 0.8$, (4.55) gives*

$$T_m = 4770°\text{K}.$$

The pressure coefficient of T_m, $\lambda_s = dT_m/dP$, is

$$\lambda_s = \frac{v_1 - v_s}{S_1 - S_s} = \frac{\Delta v^0}{\Delta S^0 - R \ln x_1},$$

since the partial molar entropy $\bar{s}_1$ in a perfect solution is $s_1^0 - R \ln x_1$. For Δv_1^0, the volume change in melting of pure iron at 3.3 Mbar, we take again Leppaluoto's estimates, $\Delta v_0 = 0.55$ cm^3/mol and $\Delta s_1^0 = 0.81$ cal/mol deg. Then $\lambda_s \cong 1 \times 10^{-3}$ °/bar. Since the pressure at the center is presently greater than the pressure at the ICB by about 0.34 Mbar, the solidus temperature at the center is $T_{mo} = 4770 + 0.34 \times 10^3 = 5110°$K.

Assuming that prior to the start of crystallization the temperature distribution was adiabatic, the temperature T_a at $r = r_i$ was initially $T_a = T_{mo} - \lambda_a \Delta P$, where

$$\lambda_a = -\left(\frac{\partial T}{\partial P}\right)_s = \frac{\alpha T}{\rho c_P}$$

---

*The same calculation at room pressure gives $T_m = 1474°$K $= 1200°$C, whereas the observed solidus temperature for $x_1 = 0.8$ is $\approx 1380°$C. This large discrepancy shows how far Fe-FeS melts depart from being perfect solutions.

and $\Delta P = 0.34$ Mbar. Taking $\alpha = 5 \times 10^{-6}/\text{deg}$, $T = 4.9 \times 10^{3}\,°\text{K}$, $\rho = 12.5$ g/cm³, and $c_P = 0.16$ cal/g deg $= 670$ J/kg deg, we find $\lambda_a = 3 \times 10^{-4}$°/bar and $T_a = 5010\,°\text{K}$. Thus at $r = r_i$ the temperature had dropped since crystallization started by $\Delta T = T_a - T_m \cong 250°$ (Figure 4-4). If this figure is typical of the whole core, the corresponding cooling rate is $\approx 2 \times 10^{-15}$°/s, assuming the inner core began to form 4 billion years ago. The rate of heat loss by the core with mass M is $Q_c = Mc_P\,\partial T/\partial t = 2.6 \times 10^{12}$ W.

This calculation omits consideration of the increase in sulfur content of the liquid caused by crystallization of iron and the corresponding lowering of the liquidus temperature. The initial sulfur content of the liquid was slightly smaller before crystallization started than it is today, and its liquidus may have been higher by some 20° or so; total cooling since the inner core began to grow would then be 270° rather than 250°. The calculation also ignores the fact that prior to separation of the inner

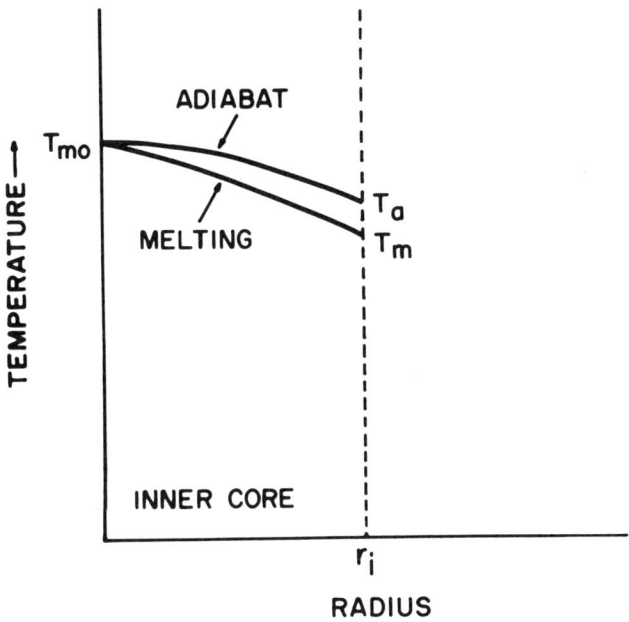

FIGURE 4-4 The temperature drop $T_a - T_m$ since crystallization of the inner core began, when the melting temperature at the center was T_{mo}. At that time the temperature T_a at the inner core boundary $r = r_i$ was on the adiabat through T_{mo}. T_m is the present temperature at the inner core boundary.

core, the pressure everywhere in the core was lower than it is today; as mentioned above, Loper (1978) estimates that the pressure at the center has risen by some 0.23 Mbar since crystallization began. The pressure difference between $r = 0$ and $r = r_i$ is also likely to have increased somewhat, since the density of the region between $r = 0$ and $r = r_i$ has also risen. Thus ΔP may have been smaller than the present value (0.34 Mbar) used here.

Finally, the latent heat released by crystallization is $\Delta h = T \Delta s = T [\Delta s^0 - Rl\eta x_1] \cong 106$ cal/g = 4.45 10^5 J/kg. For the inner core, with mass $M_i = 9.8 \times 10^{22}$ kg, $\Delta H = M_i \Delta h$ and the average rate of release Q_1 is, for an inner core 4×10^9 yr old, 3.46×10^{11} W.

These figures are, of course, very uncertain. The assumption of a perfect solution leads (see footnote, p. 96) to underestimating the solidus temperature at zero pressure by some 15 percent; if the same correction applied at core pressures, T_m would be about 5500°K, and all other temperatures would rise in proportion. The rate of cooling $\partial T/\partial t$ might not be much changed, but Δh may have been overestimated, as $\bar{s}_1$ is probably less than $s_1^0 - Rl\eta x_1$ due to the exothermic effect of mixing.

These figures differ appreciably from earlier estimates (Verhoogen, 1961), mainly because estimates of the melting temperatures of pure iron at the pressure of the ICB have greatly increased in recent years, and also because we have now considered the crystallization of iron from a FeS-Fe melt rather than from its own pure liquid.

The total rate of heat loss Q_0 of the core, assuming the inner core started to form 4 billion years ago, is

$$Q_0 = Q_c + Q_e + Q_g$$
$$= 2.6 \times 10^{12} + 0.34 \times 10^{12} + 0.66 \ 10^{12}$$
$$= 3.6 \times 10^{12} \text{ W}$$

where Q_c represents cooling of the whole core, Q_e is the latent heat of crystallization, and the third term, Q_g, comes from the gravitational energy, the largest part of which is, as we have seen, the work done by the mantle falling in on a shrinking core. The value of Q_g is, however, quite uncertain and may have been underestimated by a factor of 2 or more. The uncertainty stems mostly from our ignorance of the volumetric properties of the Fe-FeS system. Our evaluation of Q_c and Q_e was based on the perfect-solution assumption, which is almost certainly wrong.

Recall our earlier result that the radioactive dynamo requires a heat output in the range 4×10^{12}–1×10^{13} W, not markedly greater than for the gravitational dynamo. There is thus little basis for the claim that a gravitational dynamo requires a much lower heat flow into the mantle than a radiogenic one.

There is at the moment no compelling evidence to tell us that the core is not cooling and the inner core not growing (nor, for that matter, is there any evidence that the core is not heating and the inner core shrinking). If it seems more plausible to assume that the core is cooling, then surely there is a gravitational contribution to the dynamo. How large this contribution may be still seems very uncertain, mainly because of difficulties encountered in evaluating terms in Equation (4.48); these difficulties stem mostly from our ignorance of the composition of the core and of its physico-chemical properties (liquidus temperature, heat and volume of mixing, etc.). The radiogenic thermal dynamo is conceptually simpler. But who will tell us how much potassium there is in the core?

5 Core-Mantle Interactions

It was shown in the preceding chapter that the heat output of the core is likely to be somewhere between 3×10^{12} and 10^{13} W, the lower figure being the one appropriate to a cooling core with no radioactive sources. We also saw in Chapter 3 that the interpretation of mantle layer D'' as a thermal boundary layer with a steep gradient of 10°/km requires a heat flow from the core Q_c of about 9×10^{12} W if the thermal conductivity of the lower mantle is 6 W/m deg (14 mcal/cm deg s). Since it simplifies things a bit to assume a steady state, we shall henceforth assume that radioactivity of potassium is indeed the source of the core heat and assign it a value of 9×10^{12} W. For a total surface heat flow $Q_0 = 4 \times 10^{13}$ W, heat generation in the mantle must amount to 3.1×10^{13} W, corresponding to a volumetric average rate of 2.44×10^{-8} W/m^3, which is not an impossible figure, considering that uranium alone, at the accepted concentration of 18 ppb (see Chapter 2) would generate about 8×10^{-9} W/m^3. The question we now turn to is whether the heat flux Q_c from the core has any discernible effect on the behavior of the mantle. There are several possible approaches to the problem, none very successful. The first to be examined is the "efficiency" of the mantle.

EFFICIENCY OF THE MANTLE

We saw in Chapter 1 that the mantle apparently does work on continental crust, which is locally uplifted, deformed, and fractured. When

Core-Mantle Interactions 101

two continents collide to form a mountain range, the mantle on which the continents are rafted must provide the necessary mechanical energy. This is mechanical work the mantle does on its surroundings. We want to see how this mechanical energy could be related to the heat input of the core.

We consider a much simplified state in which the mantle receives heat from the core at the rate Q_c and at the temperature T_c of the core-mantle boundary (CMB). The mantle contains radioactive heat sources with density ϵ and discharges heat at its top at the rate Q_0 and at temperature T_0; in addition it does work on the continental crust (not a part of the mantle) at rate W. In the steady state, a simple energy balance requires

$$W = Q_c + \int_V \epsilon \, dV - Q_0, \tag{5.1}$$

where integration is over the volume V of the mantle. The entropy balance equation is

$$\frac{Q_0}{T_0} = \frac{Q_c}{T_c} + \int_V \frac{\epsilon dV}{T} + \int_V k \left(\frac{\nabla T}{T}\right)^2 dV + \int_V \frac{\phi}{T} \, dV, \tag{5.2}$$

where the two last terms on the right represent, respectively, the entropy generated irreversibly by heat conduction and by viscous dissipation.

Suppose that ϵ is uniformly distributed so that

$$\int_V \frac{\epsilon \, dV}{T} = \frac{\epsilon V}{T_m}, \tag{5.3}$$

where T_m, the "average" temperature of the mantle, is defined as

$$\frac{1}{T_m} = \frac{1}{V} \int_V \frac{dV}{T}. \tag{5.4}$$

Substituting in (5.1) to eliminate the unknown quantity ϵV, we get

$$W = Q_c \left(1 - \frac{T_m}{T_c}\right) + Q_0 \left(\frac{T_m}{T_0} - 1\right) - T_m J - T_m \psi, \tag{5.5}$$

where $J \equiv \int_V k \, (\nabla T/T)^2 \, dV$ and $\psi \equiv \int_V (\phi/T) \, dV$. If we make the additional assumption that $\int_V \phi/T \, dV = 1/T_m \int_V \phi \, dV = \Phi/T_m$, where T_m

is defined by (5.4) and Φ is the total viscous dissipation, we obtain for W

$$W = Q_c \left(1 - \frac{T_m}{T_c}\right) + Q_0 \left(\frac{T_m}{T_0} - 1\right) - J T_m - \Phi. \qquad (5.6)$$

Equation (5.6) is rather remarkable in that it reveals the rather special role played by Q_c. Suppose indeed that we split Q_0 into its two parts, $Q_0 = Q_c + Q'$, where Q' is the heat generated in the mantle. Then (5.6) can be written as

$$W = Q_c T_m \left(\frac{1}{T_0} - \frac{1}{T_c}\right) + Q' \left(\frac{T_m}{T_0} - 1\right) - J T_m - \Phi, \qquad (5.7)$$

and since $T_c > T_m$, the factor modifying Q_c is always larger than the factor affecting Q'. Heat from the core is more efficient than an equal amount of heat generated in the mantle. This is so because the rate of entropy production associated with core heat is less than that of mantle heat. Hence the reason for looking at mantle efficiency as a means of assessing the effect of core heat.

We suppose that the oceanic crust and lithosphere form a part of the mantle, which is permissible since they consist mainly of mantle material that participates in the mantle flow. Thus we take $Q_0 = 4 \times 10^{13}$ W and $T_0 \cong 300°K$. This is not quite correct since this value of Q_0 includes the radioactive heat generated in continents outside the mantle, while the temperature at a continental Moho, or at the base of the continental lithosphere, is certainly greater than 300°K.

To calculate T_m and J, we must define the thermal structure of the mantle. Using the mantle temperatures estimated in Chapter 3, we divide the mantle into four layers, as follows:

1. An upper boundary layer (UBL) with thickness $H = 100$ km and conductivity $k = 4$ W/m deg chosen such that the gradient β_0 at the surface is 20°/km. The gradient decreases gradually with depth to give a temperature $T_H = 1300°K$ at $z = 100$ km.

2. Layer 2, extending from 100 to 700 km, with an adiabatic temperature profile. Because of the entropy changes associated with the phase changes that occur in this layer, the adiabatic gradient is steep. The temperature scale height h_2 is taken to be 865 km, so that the temperature $T_{700} = 2600°K$. On this adiabat, the temperature at 400 km is 1838°K or 1565°C. In the layer, $k = 4$ W/m deg.

3. Layer 3, from 700 to 2800 km, with temperature scale height $h_3 = 9130$ km, to give $T_{2800} = 3300°$K. In layer 3, $k = 6$ W/m deg.

4. A lower boundary layer (LBL), extending to the core-mantle boundary, with an average gradient of 10°/km. At the core boundary, $T_{CMB} = 4300°$K and Q_c, the heat flux from the core, is 9.15×10^{12} W, if k is 6 W/m deg.

With the above numbers, the average temperature T_m comes out at 2050°K, and $J = 0.97 \times 10^{11}$ W/deg.

Substituting numbers in Equation (5.6) gives

$$W = 3.9 \times 10^{13} - \Phi \text{ W}. \tag{5.8}$$

The contribution to $W + \Phi$ of the core heat $Q_c [1 - (T_m/T_c)]$ amounts to only 0.48×10^{13} W, less than 15 percent of the total.

There is no way of estimating accurately the viscous dissipation Φ or the viscous entropy production $\psi = \int_V (\phi/T) \, dV \cong \Phi/T_m$. Practically all that can be said about Φ is that the ratio Φ/Q_0 is likely to be of the order of the ratio (d/h) of the thickness of the convecting layer to the temperature scale height h (Hewitt et al., 1975).* In our model, the ratio d/h is 0.7 in layer 2 and 0.23 in layer 3, so that, very roughly, $\Phi \cong 0.32 \, Q_0 = 1.2 \times 10^{13}$ W and W is $\approx 2.7 \times 10^{13}$ W, which is ample since W is estimated in Chapter 1 to be of the order of 10^{12} W or less.

Since Q_c contributes only a small amount to $W + \Phi$, it might be concluded that the mantle would function almost as well as a source of mechanical energy if Q_c were zero. This, however, is not a foregone conclusion. In the first place, if $Q_c = 0$, the LBL disappears, the temperature at the core boundary drops, and so does T_m, and so does W if $Q_0/T_0 > J$. Secondly, W is sensitive to J, which, in turn, is sensitive to the thermal structure of the upper boundary layer. The UBL contributes about 98 percent of the value of J because it is in this layer that temperature gradients are steepest and temperatures lowest. J is also sensitive to k, since for given Q_0, the surface gradient β_0 varies as $1/k$. A thin surface layer with much reduced conductivity $k = 1$ W/m deg might raise J sufficiently to reduce W to zero, or at least to such a low value that the contribution from Q_c would become important. Finally, the

---

*More precisely, these authors find that $\Phi = Q_0 (d/h)(1 - \mu/2)$, where μ is the ratio of internally generated heat Q' to total heat flux Q_0. Since $Q' = Q_0 - Q_c$, this gives $\Phi = 0.61 \, Q_0 \, (d/h)$.

contribution to J from horizontal temperature gradients that we have neglected may also decrease W below the value given by (5.8).

To conclude, if $W \cong 10^{12}$ W is taken as an observed quantity, it should theoretically be possible to determine Q_c from either (5.6) or (5.7). Exact calculations are, however, not feasible at the moment, mostly because of uncertainties affecting Φ, the viscous dissipation, and J, the rate of entropy production by conduction of heat, which depends only on thermal conductivity and temperature distribution. Alternatively, if both W and Q_c are chosen, the equations could be used to determine J, and from it, possible temperature distributions in the mantle.

CONVECTION PATTERNS IN THE MANTLE

Assuming now that the heat output of the core is approximately 9×10^{12} W, with a corresponding heat flux Q_c into the base of the mantle of 5.9×10^{-2} W/m², we inquire as to what the pattern of convection is most likely to be in the mantle. We note that the mantle also contains radioactive heat sources, generating on the average approximately 3.4×10^{-8} W/m³. We first examine the two heat sources separately, postponing until later a discussion of their combined effects.

CONVECTIVE PATTERN FOR DISTRIBUTED SOURCES OF HEAT

The first question that arises is whether the radioactive sources are uniformly distributed or, on the contrary, concentrated in the upper mantle. The latter hypothesis is suggested by the fact that uranium, thorium, and potassium are more abundant in the crust than in the mantle. In many crustal plutons, the concentration of those elements must decrease exponentially with depth, as shown by Lachenbruch (1968) (see Chapter 2). It has also been argued that the close-packed structure of minerals in their high-pressure form makes them inhospitable to ions as large as those of the radioactive elements, which would, so to speak, be squeezed out of the lower mantle by the high pressure prevailing there.

While all this is true, it is not clear how these elements could have been effectively eliminated from the lower mantle. Concentration in the continental crust occurred presumably by complicated and often repeated cycles of partial melting in the mantle and partial remelting in the crust. (No one seems to be very sure as to how the exponential distribution in individual plutons comes about.) Transfer of radioactive elements from the lower to the upper mantle would presumably also require at least one stage of partial melting affecting the whole of the lower mantle, not just a small blob or plume here and there.* Such partial melting and upward migration of the melt would necessarily lead to a gross chemical differentiation with respect to elements other than the radioactive ones, of which there is no sign, the whole mantle appearing on the contrary to be chemically homogeneous. Partial melting of the lower mantle requires temperatures much in excess of the present ones; it implies therefore a thermal event of the first magnitude, of which there also is no sign. Even the gravitational energy released by separation of the core would be insufficient to account for it. Thus while it is possible that the rate of production of radiogenic heat in the lower mantle could be somewhat smaller than it is in the upper mantle, the likelihood of it being zero is not great. We prefer to proceed on the assumption that the rate is uniform.

The Rayleigh number R_a appropriate for uniformly distributed sources is

$$R_a = \frac{\alpha \epsilon G R^6}{c_P \kappa^2 \nu} \tag{5.9}$$

(Chandrasekhar, 1961), where R is the outer radius of the sphere or spherical shell, G is the gravitational constant, 6.67×10^{-11} m^3/kg s^2, κ is the thermal diffusivity, ν is the kinematic viscosity, and ϵ is the radiogenic heat production rate needed to provide a total surface heat flow of 4×10^{13} W; $\epsilon = 3.4 \times 10^{-8}$ W/m^3. We take $\alpha = 2 \times 10^{-5}$/deg, $\kappa = 1.2 \times 10^{-6}$ m^2/s, $\nu = 2 \times 10^{17}$ m^2/s, and $c_P = 1.25 \times 10^3$ J/kg deg. This gives $R_a = 8.4 \times 10^9$.

What happens at such high Rayleigh number is not precisely known. Chandrasekhar (1961) showed that the critical Rayleigh number for the onset of convection is about 2×10^3 for a sphere with a free surface and 5×10^3 for a fixed-surface boundary; in both cases the onset of instability is associated with the mode $l = 1$ (in spherical harmonic language) corresponding to a single cell with one uprising current and one descending current 180° away. The critical Rayleigh number for a spherical shell remains of the same order, but the preferred mode depends now on the ratio of the outer radius to the inner radius of the shell; in circumstances similar to those obtaining in the earth, the preferred modes whould be harmonics 3 to 5, depending again on boundary conditions. (The harmonic $l = 3$ means that the convection pattern in

---

*How pervasive the process must be to remove all radioactive elements from the lower mantle may be gauged by noticing that in spite of its long geological history of repeated local partial melting, the upper mantle still contains appreciable radiogenic heat sources.

the mantle breaks down into three cells, rather than just one as in the case $l = 1$).

Hsui *et al.* (1972) have numerically pursued the problem for a sphere to higher Rayleigh numbers. For free-surface boundary conditions, the single-cell pattern is maintained up to a Rayleigh number of 5×10^6; for a fixed-surface boundary, a two-cell pattern appears at $R_a = 5 \times 10^5$. What happens at $R_a \cong 10^{10}$ for a spherical shell like the mantle can only be guessed. Presumably, the high Rayleigh number and finite thickness of the shell will both tend to break the circulation into smaller cells, with harmonic number higher than 5.

There is a vast literature on convection patterns in the plane geometry of an infinite layer, but most of it refers to the case of a fluid heated from below rather than to a fluid heated from within. The adaptation to the geometry of a spherical shell is not straightforward. In particular, it is not clear how the elongated cylindrical rolls with horizontal axis characteristic of convection in a plane layer could survive transposition to a thick shell. Examination of this problem requires close attention to the nonlinear terms in the dynamic equations. Busse (1975a) finds that a spherical geometry leads to a qualitative difference between convective patterns of odd and even harmonic order l; it effectively prohibits patterns of odd order.* This difference does not have a direct analogue in the case of a plane layer, for which solutions with different wave numbers differ only quantitatively. A possible pattern of even order is shown in Figure 5–1. Busse points out that his results are apparently not applicable to the mantle, since gravity and other geophysical data for the earth do not seem to show a preference for even orders, a fact he attributes tentatively to the inhomogeneous structure of the upper mantle and continents. We may also note in passing an interesting attempt by Walzer (1971) to determine possible convection patterns from a purely kinematic argument based on group theory. Assuming at the start that flow lines are circular, so as to minimize dissipation, he seeks the patterns of superposed tiers of convective cells that have maximum symmetry or regularity (in the sense of group theory) and that lead to a maximum rate of outward heat transport.

The facts of plate tectonics strongly suggest that the convection pattern in the mantle depends on time; plates suddenly change direction, oceanic ridges apparently migrate in longitude or latitude, or both, and large plates break up into smaller ones. Whereas it is generally

---

*A spherical harmonic function is such that its value at opposite points on the sphere is equal and of the same sign for even orders, of opposite sign for odd orders.

Core-Mantle Interactions 107

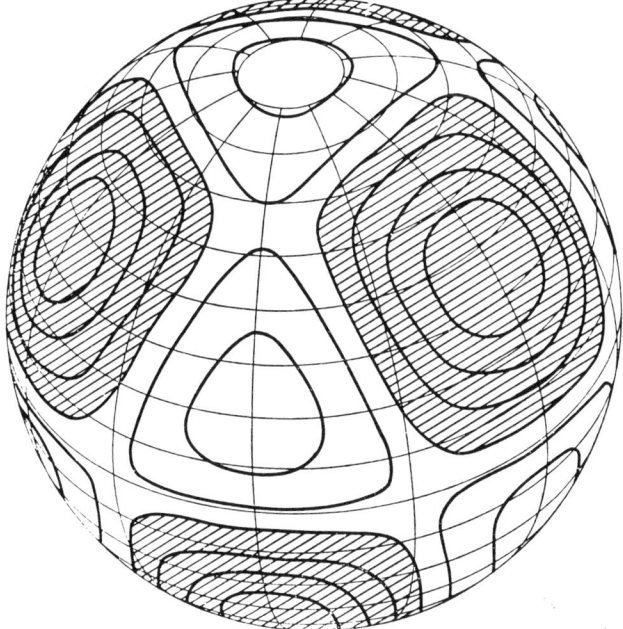

FIGURE 5-1 A possible pattern in a spherical shell with internal heat sources. The motion is ascending or descending in the shaded area according to circumstances. Reproduced with permission from Busse (1975a).

accepted that flow is steady at the onset of instability in fluids with high Prandtl number, there is experimental evidence (e.g., Krishnamurti, 1970) that the flow may become time dependent at Rayleigh numbers much greater than the critical R_c characteristic of the onset of instability. The onset of time dependency depends on both the Rayleigh number R_a and the Prandtl number $P_r = \nu/\kappa$, and, as Jones (1977) points out, there is no experimental evidence pertaining to fluids with very high P_r and small R_a/P_r ratio, as obtains in the mantle. The importance of the latter condition ($R_a/P_r \ll 1$) has been pointed out by G. M. Corcos (personal communication to Jones, 1977). If we nevertheless extrapolate Krishnamurti's results (Figure 5-2) to large P_r, it appears that in a fluid heated from below, the flow might become time dependent at $R_a > 10^5 - 10^6$. Numerical calculations at infinite P_r (Busse, 1967) show that as the Rayleigh number is increased above its critical value, the pattern of flow changes from a two-dimensional (elongated rolls) to a three-dimensional, but steady, pattern that, at a still higher Rayleigh

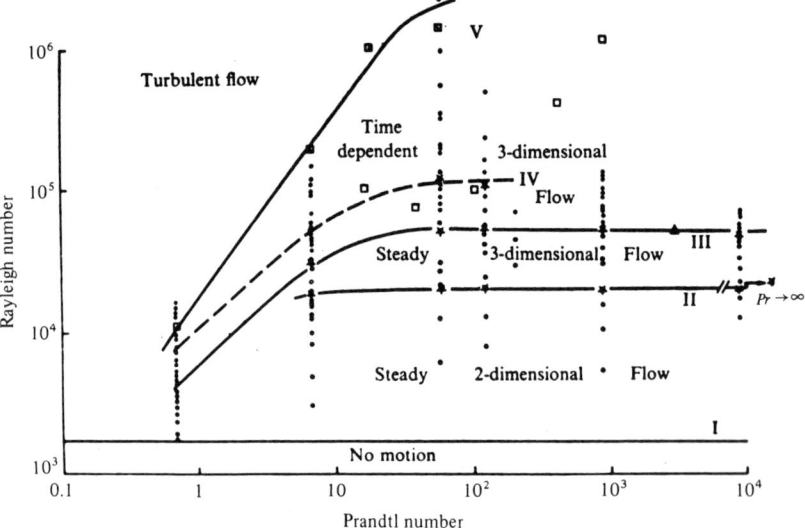

FIGURE 5-2 Flow regimes in a layer of fluid heated from below. The diagram illustrates the dependence on Rayleigh and Prandtl numbers of the onset of time-dependent flow. Reproduced with permission from Krishnamurti (1970).

number, becomes time dependent. This time-dependent flow is quite different from the oscillatory motion that appears in fluids (e.g., mercury) with very low Prandtl number, and that can be explained with linear theory. The time dependence we are concerned with here, and which results from the nonlinear terms in the dynamic and heat equations, consists of a slow change in pattern, well illustrated by the numerical experiments of McKenzie *et al.* (1974). In one particular experiment (see their Figure 17), a fluid heated from within is enclosed in a rectangular box twice as wide as it is deep. The Rayleigh number is 1.4×10^6. The motion, which initially consists of a single cell or roll, is unstable and gradually breaks down into two rolls with both ascending plumes at the vertical boundaries of the box. This is followed by formation of four rolls, two of which are larger than the others. The larger rolls grow at the expense of the smaller ones until the two-roll pattern is approximately restored. A new instability in the upper boundary layer develops, leading again to the four-roll pattern, and the cycle is repeated on a time scale that, for the mantle, would be on the order of 10^7–10^8 yr. The authors have reason to believe that this unsteady flow is a feature of the solution of the nonlinear equations rather than an artifact of the numerical scheme used in the calculations.

We need to mention one more feature of the convection pattern in a fluid heated from within to which McKenzie has drawn attention (McKenzie *et al.*, 1974). In contrast to fluids heated from below, in which upwellings and downwellings are localized and symmetrical, the pattern in fluids heated from within tends to consist of localized downwellings separated by regions of diffuse and slow upwelling (Figure 5–3). This occurs because, if heat is generated everywhere, the flow must bring all parts of the fluid close to the upper surface to permit them to lose heat by conduction. Also, the buoyancy force is distributed

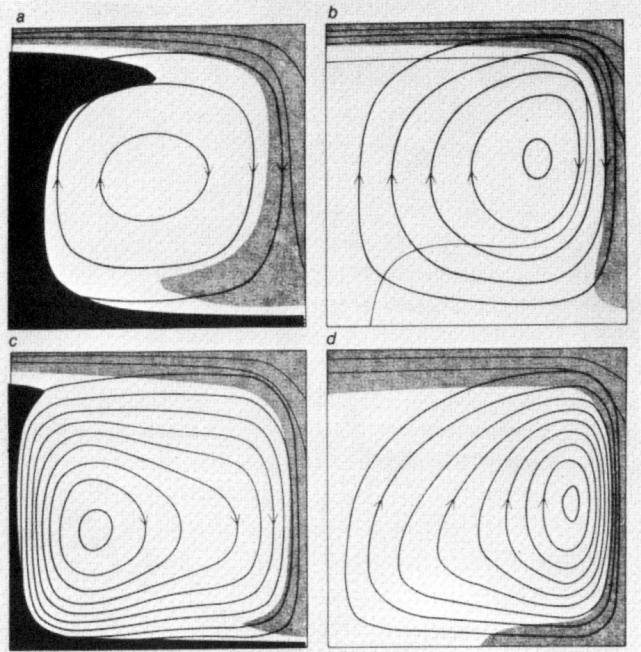

FIGURE 5–3 Computer simulation of convection cells. The two left cells, (a) and (c), are for liquid heated from below; those on the right, (b) and (d), have internal heat sources. The two top patterns, (a) and (b), are for constant viscosity; in (c) and (d) the viscosity decreases with increasing temperature. In case (a), the pattern is nearly symmetrical, whereas in (b) there is a relatively narrow sinking sheet with upwelling everywhere else. Variable viscosity does not much alter the pattern, (d), for a fluid heated from within; heating from below, as in (c), produces a hot rising sheet narrower than the cold sinking sheet. Deep black is hot; gray is cold. Reproduced, with permission from D. P. McKenzie and F. Richter, *Scientific American*, 235, 82, 1976. Copyright © Scientific American.

throughout the fluid rather than concentrated in the lowermost layer as in the case of a fluid heated from below. This raises the question as to whether localized upwelling such as is believed to occur at oceanic ridges or in deep-mantle plumes could ever develop in a fluid with only internal heat sources (McKenzie and Weiss, 1975).

HEAT FROM THE CORE

We now examine what happens to the mantle when it receives from the core, through its lower boundary, a heat flux $q = Q_c/S = 5.9 \times 10^{-2}$ W/m² (S is the area of the core-mantle boundary).

The appropriate Rayleigh number based on q rather than on an assumed constant temperature difference between top and bottom of the mantle is (Foster, 1971)

$$R_q = \frac{\alpha g\, q\, d^4}{\kappa^2\, \nu \rho\, c_P}, \qquad (5.10)$$

where d is the thickness of the layer. Using the same numbers as before, we get $R_q = 5.7 \times 10^8$. For an incompressible fluid with infinite Prandtl number, in the Boussinesq approximation, Foster shows that for $R_q > 10^7$ the flow is intermittent with a mean period of intermittency τ given by

$$\tau = \frac{\nu \rho\, c_P}{\alpha g\, q} \qquad (5.11)$$

and a horizontal wave length

$$a = \frac{2\pi}{0.13} \left(\frac{\kappa^2 \nu \rho\, c_P}{\alpha g\, q} \right)^{1/4}. \qquad (5.12)$$

The mechanism that leads to intermittency may be the one illustrated by Howard (1966), and further explored by Jones (1977) and by Elsasser *et al.* (1979). At the start, a thermal conductive boundary layer forms at the CMB to carry the heat q into the mantle. Initially it is so thin that the local Rayleigh number, based on the temperature difference across the boundary layer and its thickness δ, is less than about 10^3 and too small to induce convection; but as the thickness of the boundary layer increases it becomes unstable and ejects matter, which detaches itself from the CMB and rises by buoyancy. It is then replaced by a cold descending layer that in turn heats up and rises. Jones, using numbers slightly different from those used here, calculates that for $\nu = 10^{17}$ m²/s, τ is about 100 million years and the horizontal wavelength is about 700 km.

The mechanism described above that leads to a time-dependent behavior is essentially the same as the one analyzed by McKenzie *et al.* (1974). It stems from the instability of a thermal boundary layer in which a steep conductive temperature gradient develops. As the conductive layer thickens, the local Rayleigh number in that layer exceeds the critical value, and the layer becomes unstable. Foster (1971) predicts that this will happen at $R_q > 10^7$, but McKenzie *et al.* point out that this number (10^7) is an artifact of Foster's numerical scheme, which considers only temperatures averaged over a horizontal plane; according to them, the flow will become time dependent at a lower value of the Rayleigh number.

Foster's calculations, in plane geometry, assume q to be uniform, which it will not be in the geometry of the core. The heat flux q, it will be remembered, is the heat transported, mostly by convection, toward the surface of the outer core; its value at a point on the CMB will depend on the pattern of core convection. The inner core plays a role here, as Busse has shown (see, for instance, Busse and Cuong, 1977). Because of the influence of rotation, the convective regime will be different inside and outside the cylindrical surface tangent to the inner core surface at its equator; there are regions inside and near the cylindrical surface where the radial heat transport becomes negative near the CMB. A similar dependence of heat flux on latitude is also apparent in Gilman's (1977) study of convection in a rotating spherical shell. Gilman finds that in general the heat flux is greatest in a relatively narrow belt around the equator, with a minor peak in polar regions and a minimum in midlatitudes. Details of the pattern turn out to be quite sensitive to the Rayleigh number. As Gilman's study was designed with the sun in mind, he chose a value of 10^5 for the Taylor number $T = 4\Omega^2 d^4/\nu^2$ (Ω is the angular velocity), $P_r = 1$, and Rayleigh numbers in the range 1×10^4–4×10^4. The numbers relevant to the core are rather different ($T \cong R_a \cong 10^{29}$), yet the ratio $F^2 = P_r T/R_a$ is about the same for the core and in Gilman's analysis, and the effect of rotation on the latitudinal dependence of the surface heat flux should be about the same in the core and in the sun. There is thus reason to believe that the heat flux across the CMB will not be uniform; it will probably be maximum at the core's equator and also, according to Gilman, fluctuate appreciably in longitude. It is indeed a characteristic of rotating shells that the flow is not axisymmetric.

112 ENERGETICS OF THE EARTH

The effect of the nonuniform q on the intermittent convection in the mantle has not been worked out.

ADDITIONAL FACTORS IN CONTROLLING THE FLOW

To this already sufficiently complicated picture we must now add several factors not considered so far. These are (1) possible temperature dependence of some physical properties of the system, (2) von Zeipel instability, (3) the effect of surface plates, and (4) the effect of phase transitions.

TEMPERATURE DEPENDENCE OF PHYSICAL PROPERTIES

From early Bénard experiments on thin layers of fluid heated from below emerged a classical pattern of hexagonal cells that has long been considered typical. It is now known that at low supercritical Rayleigh numbers the stable pattern consists of two-dimensional rolls, not of three-dimensional hexagons; these turn out to be the effect of the temperature dependence of the surface tension of the fluid used in the experiments (for a review, see Palm, 1975). The nonlinear mechanism that selects the hexagonal cellular flow in convection driven by surface tension also operates in a fluid with temperature-dependent viscosity, or with temperature-dependent thermal diffusivity (Palm, 1975). In the latter case, the variable κ implies a nonlinear conduction temperature profile, as could be produced also by distributed internal heat sources. Figure 5–3 illustrates the effect of variable viscosity on convection in a layer of fluid heated from below.

THE VON ZEIPEL INSTABILITY

The von Zeipel instability arises from von Zeipel's theorem, which states that hydrostatic equilibrium can be achieved in a homogeneous, rotating, self-gravitating body with distributed internal heat sources only if ϵ is distributed according to a very special law (Verhoogen, 1948). It is well known that hydrostatic equilibrium in a homogeneous body requires that density, pressure, and temperature be constant on surfaces of constant potential (gravitational plus centrifugal). For angular velocity Ω the potential U satisfied

$$\nabla^2 U = -4\pi G\rho + 2\Omega^2, \qquad (5.13)$$

whereas in the steady state the temperature satisfies

$$\nabla^2 T = -\epsilon/k. \tag{5.14}$$

Clearly, a relation among ϵ, ρ, and Ω is implied by the condition that equipotential and isothermal surfaces must coincide. This relation is (Eddington, 1926)

$$\epsilon_m = 4\pi G f \left(1 - \frac{\Omega^2}{2\pi G \rho}\right), \tag{5.15}$$

where $\epsilon_m = \rho\epsilon$ is the heat source density and f is such that

$$f = q/g = \text{constant}. \tag{5.16}$$

Equation (5.14) implies that the ratio of the heat flux q to gravitational acceleration g must be constant throughout the body. When conditions (5.15) and (5.16) are not satisfied, the equipotential surfaces that are flattened by the rotation no longer coincide with isothermal surfaces, the density varies along equipotential surfaces, and $\nabla\rho \times \nabla U \neq 0$. Thus hydrostatic equilibrium ($\nabla P = \rho \nabla U$, hence $\nabla\rho \times \nabla U = 0$) becomes impossible.

The character of the ensuing motion depends on the properties of the fluid. When the Prandtl number and viscosity are low, as in the outer core, only the Coriolis force can balance the von Zeipel term $\nabla\rho \times \nabla U$, and a geostrophic flow with nonuniform rotation will presumably develop (this problem has not been solved exactly yet). In the highly viscous mantle, the Coriolis force is negligible, leaving only the viscous force to balance the von Zeipel buoyancy. The problem has been investigated by McKenzie (1968), who wanted to find out, among other things, if the von Zeipel flow, also known as the Eddington current, could support the nonhydrostatic part of the equatorial bulge. He examined both compressible and incompressible spheres. In both cases, the flow takes place in meridional planes, with currents rising at the poles and sinking at the equator in the incompressible case; the velocities in this case are very small. Rather surprisingly, McKenzie found that compressibility reverses the sense of the circulation and increases the velocity by a factor of 30. This, however, might be a result of his erroneous assumption that the volumetric heat generation rate ϵ remains uniform in the compressible case; it is, of course, $\epsilon_m = \rho\epsilon$ that remains constant. Furthermore, McKenzie's analysis applies only if the viscosity is extremely large ($>10^{27}$), in which case there would be no convection of any kind. Thus the problem remains open. McKenzie's principal argument for disregarding Eddington currents is the geographic orientation of orogenic belts and midoceanic ridges, which show complete disregard for the rotational axis, in contrast to the axisymmetric Eddington currents and their north-south (or south-north) surface velocities.

No one will argue that von Zeipel's effect is the only cause of convection in the mantle, any more than one would argue that the small climatic difference in temperature between the poles and equator on the surface of the earth has important mechanical effects on mantle circulation. It is important to remember, though, that this effect amounts to an inherent instability at zero Rayleigh number that could well trigger other near-marginal instabilities and help select among modes with nearly the same critical Rayleigh number.

EFFECTS OF SURFACE PLATES

Suboceanic upper mantle differs from subcontinental upper mantle with respect to distribution of radioactivity and temperature and with respect to rheological and mechanical properties. Plates carrying continents seem difficult to move around; they certainly move more slowly than purely oceanic plates. Because of their buoyancy, continents cannot be subducted; when they collide to form mountains, kinetic energy of mantle flow is transformed to strain and potential energy outside the mantle, and consequently the mantle flow must slow down. It is clear that constraints on the motion of plates will have an effect on mantle flow, extending perhaps throughout the mantle (Davies, 1977). Much of the time-dependent behavior of plates may be due to such surficial causes rather than to intrinsic properties of the convecting system. If this is true, the plate motions at any time depend on what has happened before; to account for the present-day pattern may well require knowledge of the events of the past billion years.

EFFECTS OF PHASE TRANSITIONS IN THE MANTLE

The problem is to find in what way the existence of phase transitions in the mantle, notably near 400-km and 650-km depth, could affect the pattern of convection. Can a rising or sinking current go through such a transition, or will the flow be restricted to homogeneous layers lying between phase discontinuities?

In spite of much work (Verhoogen, 1965; Schubert and Turcotte, 1971; Richter, 1973; Schubert *et al.*, 1975) the question has not been completely answered yet. Schubert and Turcotte performed a stability analysis for a fluid layer heated from below in which a horizontal boundary $z = 0$ separates the two phases. The phase transition is assumed to be univariant, i.e., at any pressure the equilibrium temperature T_e is determined and falls on a Clapeyron curve with slope $dT_e/dP = \Delta V/\Delta S = T_e \Delta V/\Delta H$, where ΔV and ΔS are respectively the volume and entropy changes associated with the phase transformation and $\Delta H = T_e \Delta S$ is the latent heat of the transformation. Schubert and Turcotte show that the critical Rayleigh number for convection through the phase boundary depends now on two additional numbers, S and R_Q, the first of which is a measure of the density jump at the transition, the second of which measures the latent heat ΔH. They find that for given R_Q, the critical Rayleigh number decreases with increasing S; for given S, the critical Rayleigh number increases with increasing R_Q. Their analysis is, however, marred by their choice of an erroneous boundary condition, namely, that at the phase boundary $z = 0$ the vertical component of the velocity is continuous, $w_1 = w_2$. Conservation of mass requires, of course, that $\rho_1 w_1 = \rho_2 w_2$, so that if $\rho_1 \neq \rho_2$ (as must necessarily be the case if there is a phase transition), w_1 cannot be equal to w_2.

In 1975, Schubert et al. extended the analysis to the case, more relevant to the mantle, of a divariant phase transition such as the olivine-spinel transition. The transition is said to be divariant because the system now has two components (e.g., Mg_2SiO_4 and Fe_2SiO_4). There are now four variables (not three, as Schubert et al. assert), namely, P, T, and the mole fractions of one of the two components in both phases. The effect of this is that the equilibrium temperature (or pressure) is no longer determined by the pressure (or temperature) as it would be in an univariant equilibrium. If pressure is raised at constant temperature, the transformation of olivine into spinel starts at a lower pressure, P_1, and ends at a higher pressure, P_2; within the range $\Delta P = P_2 - P_1$ both phases coexist, with different and varying compositions. It is important to remember that the composition of these phases also changes as a function of pressure. As the pressure rises, the yet untransformed olivine becomes progressively richer in magnesium, as does the growing spinel, which, at the beginning of the transformation, is richer in iron than the original olivine. Similarly, at constant pressure the transition occurs with rising temperature in the sense spinel $\rightarrow$ olivine, and is spread over a temperature range ΔT. On a P-T plot (Figure 5-4) for a given initial gross composition, the beginning and end of the transformation are indicated by the curves 1 and 2; for any point lying between these curves, the two phases coexist.

The first step taken by Schubert et al. (1975) is to determine the adiabatic (or better, isentropic) gradient within the divariant zone, i.e.,

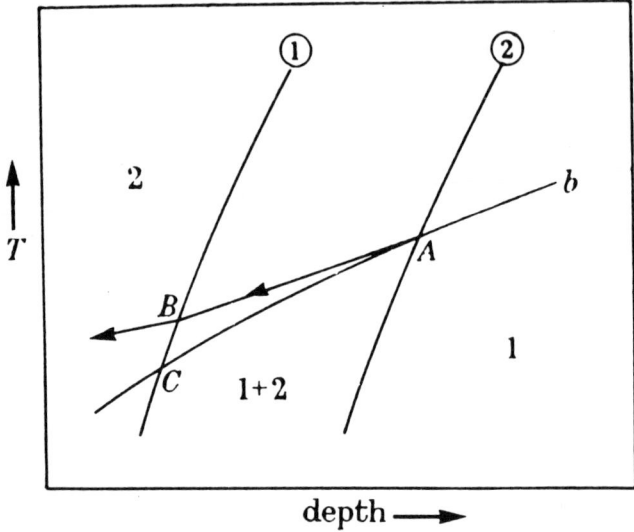

FIGURE 5–4 Convection through a divariant phase transformation. Curves (1) and (2) represent, respectively, the beginning and end of the transformation from phase 1 to phase 2. AC is the temperature distribution prior to the onset of convection, AB is an adiabatic path, starting at A, through the divariant zone.

to define the slope of the curve AB in Figure 5–4 on which the entropy at any point is equal to the entropy at the starting point A. Because the compositions of both phases vary from point to point along the curve, an exact calculation is difficult. Schubert *et al.* make simplifications that amount to reducing to zero the width of the divariant zone. The general trend of their analysis is, nevertheless, correct. The two-phase region may, to a first approximation, be treated as homogeneous, with an effective expansion $\bar{\alpha}$,

$$\bar{\alpha} = \alpha - \frac{\Delta \rho}{\rho \Delta T},$$

where α is the ordinary thermal expansion of olivine or spinel (assumed to be equal); the second term represents the effect of the transition with total density jump $\Delta \rho$ taking place over a temperature interval ΔT, at constant pressure, and is very much larger, perhaps 100 times larger, than α, which can be safely dropped. (Note that for the olivine-spinel transition $\Delta\rho$ is <0 for $\Delta T > 0$.) Similarly, the effective heat capacity $\bar{c}_P$ is defined as

$$\bar{c}_P = c_P + \frac{\Delta H}{\Delta T} = c_P + \frac{T\Delta S}{\Delta T}.$$

The adiabatic pressure coefficient of temperature is

$$\left(\frac{\partial T}{\partial P}\right)_s = \frac{\bar{\alpha}T}{\rho\bar{c}_P} = \frac{-\Delta\rho}{\rho^2(c_P\Delta T + T\Delta S)}. \tag{5.17}$$

If $\Delta T \to 0$, as for a univariant transformation, the right side of (5.17) reduces to $\Delta V/\Delta S$ (where $V = 1/\rho$); the adiabatic slope is then equal to the slope of the Clapeyron curve. For the olivine-spinel transition, $T\Delta S$ is of the same order as, or somewhat larger than, $c_P\Delta T$, so that the isentropic slope will be approximately one half to one fourth of the slope of the curves bounding the divariant region.

Schubert et al. (1975) then proceed to make a stability analysis of a system consisting of two superposed layers separated by a divariant two-phase layer of thickness $2d$; the total thickness of the three layers is $2D$. A parameter R is defined as

$$R = \frac{\bar{\alpha}(\bar{\beta}_a - \beta_a)gD^4}{\kappa\nu},$$

where $\bar{\alpha}$ and $\bar{\beta}_a$ are, respectively, the thermal expansion coefficient and the adiabatic gradient in the divariant region, while β_a is the adiabatic gradient in the single-phase layers. The critical Rayleigh number R_c now depends on the ratios d/D and $\bar{\alpha}/\alpha$ and on R; it rises rapidly when R increases from 10^2 to 10^6 (for $d/D = 0.05$, $\bar{\alpha}/\alpha = 100$). The analysis, however, is carried out on the assumption that the Boussinesq approximation is valid in all three layers, including the divariant one. This is a surprising assumption to make, considering that the very essence of the two-phase region is its marked density gradient; a particle of fluid moving through it sustains over a short vertical distance a change in density of the order of 10 percent of the density itself. It is generally recognized (Hewitt et al., 1975) that the Boussinesq approximation is valid only if the ratio d/H_T of the thickness of the convecting layer to the temperature scale height $H_T = c_P/g\alpha$ is < 1.0. The condition is generally met in the mantle, where $H_T \cong 10^6$–10^7 m, but it may not be met in a two-phase region where $H_T \sim 10^4$–10^5 m.

However that may be, the main result of Schubert *et al.* (1975) is that the stability of the mantle is not changed much if the phase change is divariant rather than univariant. As applied to the olivine-spinel transition, their calculations seem to show that for a mantle viscosity less than $3 \times 10^{22} - 1 \times 10^{23}$ cm²/s, double-cell convection requires a smaller overall superadiabatic gradient than single-cell convection through the phase change region ("double cell" means convection confined to the two layers above and below the transition zone, with no flow across it). For $10^{23} < \nu < 10^{24}$, convection through the phase transition is the preferred mode. On the other hand, Richter (1973) finds, again by taking the olivine-spinel transition as a model, that (1) the vertical scale of motion is the entire depth of the fluid, (2) the horizontal scale is not significantly changed from the case of a single-phase fluid, (3) the amplitude of the motion is not significantly changed, as the buoyancy effects are largely balanced by the effect of latent heat, and (4) the phase boundary varies in depth by as much as 30 km, for vertical velocities of the order of 10^{-1} cm/yr.

Nothing can be said of the effect on convection of the phase transition near 650 km since its mineralogical nature is not definitely known; even the sign of the entropy change ΔS is unknown. McKenzie and Weiss (1975), however, argue that this phase change is important because of differences in mechanical properties of the two phases, or phase assemblages. These differences will lead, according to the authors, to formation of a mechanical boundary layer across which heat is transferred by conduction; whether this boundary layer breaks up into plates or not depends on whether the convective forces exerted on it from below are sufficient to overcome its strength. There is, of course, as yet no compelling evidence for notable differences in mechanical properties of the upper and lower mantles. Thus McKenzie and Weiss's argument appears too speculative to conclude from it that upper and lower mantles convect as two separate units, with no flow across the boundary.

INFLUENCE OF THE CORE ON CONVECTION IN THE MANTLE

From the lengthy discussion of the preceding sections, it must be clear that the convectional pattern of the mantle is presently unknown. Neither observations nor numerical experiments suffice as yet to describe accurately what may be happening. The only point that seems relatively clear is that convection must affect the whole mantle; because of relatively high Rayleigh numbers, the pattern of flow is presumably a complicated, three-dimensional, time-dependent one proceeding simultaneously on several time, length, and height scales.

A few years ago, the author was inclined to think that the core exerts a major influence on the mantle, that without core heat, convection in the mantle would be less vigorous than it is and possibly restricted to its upper part. This is not so clear anymore, for it is now certain, from the energy balance, that a large fraction of the earth's heat must come from the mantle. Since removal of all radioactivity from the lower mantle would require a major episode of melting and would produce a marked petrological differentiation that is not observed, it must be assumed that, at least to a first approximation, radioactivity is more or less evenly distributed in the mantle at a concentration corresponding to a supercritical Rayleigh number. The mantle is unstable, with or without the core. It will thus be difficult to pin down, in the largely unknown pattern of mantle convection, what may or may not be specifically related to an influx of core heat.

There are several ways by which, as discussed above, it might be possible to detect and single out the effects of core heat. The first was mentioned in the first section of this chapter, which dealt with the amount of mechanical work the mantle can do on continental plates. But as we saw, no conclusion can be drawn as yet, mostly because of the large uncertainties affecting the magnitude of the viscous dissipation function and the temperature distribution.

A second criterion for input of heat from the core was thought to be the time-dependent character of mantle convection, which could easily be explained on the Foster-Jones model of a mantle vigorously heated from below. It now seems likely that the Rayleigh number for radiogenic sources alone would be sufficiently high to induce time dependency.

Finally, it was shown that, because of rotation, heat in the core will be preferentially transported to equatorial and polar regions, with a minimum in midlatitudes. No such effect is clearly recognizable on the earth's surface. But it is a long way from the core to the surface, and the amplitude of any horizontal temperature variation on the CMB is likely to be markedly reduced, by conduction and convection, before it reaches the surface. Since the heat flow from the core represents, at most, less than one third of the surface heat flow, it is not likely that the effect could be readily detected on the surface.

There is, however, one feature that may help to decide if core heat plays a role in shaping the convection pattern in the mantle. We noted earlier that, as pointed out by McKenzie *et al.* (1974), when heat sources are distributed, upwellings tend to be diffuse and spread out over most of the convective volume. Upwelling in the mantle, on the contrary, seems concentrated in narrow sheets (ridges) or thin plumes, a feature more commonly observed in numerical experiments on fluids heated from below. The argument is, however, not decisive, because it has not been shown that the observed pattern of high heat flow at narrow oceanic ridges could not also be accounted for, in the whole or in part, by other factors such as the mechanical properties of the plates, the temperature dependence of viscosity, and so forth.

INFLUENCE OF THE MANTLE ON CONVECTION IN THE CORE

Imagine a core entirely devoid of heat sources so that it cannot convect. Imagine also a lower mantle with radiogenic sources sufficient to cause convection in it. This convection would presumably lead to the formation, on the CMB, of cold spots corresponding to localized downwellings in the mantle. Could the horizontal temperature variations on the CMB cause convection in the core sufficient to generate a magnetic field?

The question has not, to the writer's knowledge, been examined in detail; but a preliminary answer seems to be no (Jones, 1977), as the motion is likely to be restricted to a very thin zone at the very top of the core. There is no likelihood, in particular, of the development of the differential rotation necessary for the growth of a powerful toroidal field.

If, however, the core is unstable, while the mantle convects intermittently, it seems likely that the pattern of convection in the core will reflect the temperature variations at the CMB. These variations include an overall decrease in temperature at the CMB during the convective part of the mantle cycle; hence, presumably, the Rayleigh number for the core will increase, and the core convective pattern will intensify and change. Should reversals of the magnetic field be caused by changes in the latitude of the zone of cyclonic circulation, as suggested by Levy (1972), then possibly a change in temperature on the CMB might induce a change in the polarity of the field, or in the mean frequency of reversals. This is what prompts Jones (1977) to argue that the estimated period of intermittency of mantle convection could be related to long-period changes in the average rate of magnetic reversals.

On the whole, it does not seem impossible that long-period ($>10^7$ yr) fluctuations of the magnetic field might correlate with changes occurring in the mantle and at the CMB. It is difficult, however, to prove such a correlation because of the long time delay ($\approx 10^8$ yr?) between events in the core and their manifestation at the surface. A few years ago, Irving and Park (1972) called attention to the apparent paleomagnetic polar wander path for North America, which occasionally shows striking changes in direction. Since the apparent polar wander path reflects mostly changes in the motion of the plate from which it is determined, these "hairpins" in the polar wander path imply sudden changes in the motion of the plate, which, so to speak, stops and then starts back nearly in the direction from which it came. It is a curious fact that the two latest hairpins happened during the two latest periods (Cretaceous and Permian) in which the field remained in the same polarity for a time on the order of 30 million years. But this can hardly be more than a coincidence, since it is hard to see how a change in plate motion could affect instantaneously the convection pattern of the core, or vice versa.

Thus, it would seem that at this stage of the game there is not much definite evidence for interaction between mantle and core, with one minor exception. This exception refers to the so-called irregular fluctuations in the length of the day, which almost certainly arise from the electromagnetic torque exerted by the core on the mantle. But this is not a geophysical phenomenon of the first magnitude, as far as energy is concerned. The very existence of the magnetic field implies, as we have seen, that the core must be losing heat to the mantle at a rate of between one tenth and one quarter of the total heat loss from the earth; yet there is no way of proving in the present state of the art that the core heat plays a significant role in the energetic economy of the mantle, or that without it plate tectonics would not be operational or would operate in a very different mode. I, for one, suspect that core heat is significant, but I must admit that I cannot prove it.

SOME FINAL REMARKS

Let us now return to some of the broad questions raised in the introduction to Chapter 1 and summarize some of the evidence. We consider first the matter of possible energy sources, the two contenders being radioactivity and gravitational energy. With regard to radioactivity, we distinguish between the "low potassium" and "high potassium" models. The first assumes the abundance of radioactive elements to be that given in Table 2–1; the high-potassium model assumes that the abundance of potassium in the earth is close to its solar abundance or to its abundance in chondritic meteorites, with much of the terrestrial potassium buried in the core. The latter model seems preferable, if only because it removes the obligation of explaining why the earth should be more depleted in potassium than chondritic meteorites. The low-potassium model produces today somewhat less heat than now escapes at the surface; it implies therefore a cooling earth.

The magnitude of the gravitational energy source is much less certain. We include in it (1) that part of the gravitational energy released during accretion that was not immediately radiated away (what we first called "original heat," (2) the gravitational energy released by separation of the core, (3) the gravitational energy released by separation of the inner core, (4) the tidal dissipation that may have been important during the early days of the earth-moon system, and (5) the kinetic energy of impacting meteorites of large size traveling at orbital velocities. Only two of these quantities (separation of core and separation of inner core) are known, if only rather approximately. The other three items (original heat, tidal dissipation, and meteorite impact) are very uncertain and highly speculative. Separation of the inner core is not an important item in the earth's energy budget. It implies a cooling earth and is therefore compatible only with the low-potassium model. Separation of core from mantle, if it ever took place, must have been an energetic event of the first magnitude ($\approx 10^{31}$ J), and so must have been the first item in the above list. All told, it would not take much stretching of the imagination to suggest that all gravitational sources combined could indeed provide all the energy needed to run the earth. Since we know for certain that the earth also contains some radioactive elements, it seems that there is no problem in finding adequate sources; most probably radiogenic and gravitational sources are both involved. But in what proportion?

Radioactive and gravitational sources are both time dependent, the gravitational ones strikingly so. It may then be surprising that much of our previous discussion dealt with steady-state models, e.g., the efficiency of a steady-state mantle or a steady-state core. This emphasis on the steady state is not purely for the sake of mathematical simplicity. It also stems from an analysis by Tozer (1972, 1977), who argues that in a convecting planetary body in which viscosity depends strongly on temperature, the temperature distribution is in the long run nearly independent of the intensity of the heat sources. This happens because an increase in heat generation that will raise locally the temperature will also lower the viscosity, thus allowing for a more vigorous convection that will rapidly carry the excess heat away; if the heat sources decrease in intensity, the viscosity will rise, convection will slow down, less heat will be carried away, and the temperature will rise to near its previous value. This self-regulating property of a convecting planet adds credibility to the steady-state hypothesis.

The high-potassium model is essentially a steady-state model. It provides, as we have seen, all the heat now escaping from the earth, and is therefore incompatible with cooling. It should perhaps better be called quasi-steady, since radioactive sources necessarily decrease with time; yet a rough equality of surface heat flow and radioactive heat generation may have prevailed throughout most of the earth's history, while mantle temperatures remained nearly constant by the operation of Tozer's self-regulating mechanism. The high-potassium model does not, however, exclude early contributions from gravitational sources. As noted in Chapter 3, present temperatures in the lower mantle and core seem to be several thousand degrees higher than the original temperatures—for instance, those calculated from accretion theory by Hanks and Anderson (1969). Even if the core had retained all the heat generated in it by its 0.1 percent potassium, its temperature could not have risen by more than some 700° in 4 billion years. Thus it seems that a gravitational source (e.g., separation of the core) may be needed to account for present-day temperatures. We recall that separation of the core would release about 10^{31} J, enough to heat the whole core by 7500°, or the whole earth by 1500°.

The low-potassium model, on the other hand, is incompatible with any kind of steady state since it implies that the mantle must be cooling at a rate of approximately 100° per 10^9 yr in order to provide for the difference between present heat flow and radiogenic heat production (see Table 2–1); the core must also be cooling to provide the gravitational energy that drives the dynamo. It is not certain that separation of the core could provide enough energy to heat up both the mantle and core to the somewhat higher temperatures required in this model; possibly a large tidal event, or a higher initial accretion temperature, will be needed. At any rate, the low-potassium model demands more gravitational energy than the high-potassium model.

Significant as the release of gravitational energy may be, sight should not be lost of the part played by the gravitational field in the development of the "structures" alluded to in the introduction to Chapter 1. As mentioned, our problem is not only to find adequate energy sources; we must also account for observed structures. Regional variations in heat flow as a function of distance from an oceanic ridge, or as a function of geological age of a continental province, are examples of such structures, as are also horizontal temperature differences in the upper mantle, localization of volcanoes, or, more generally, any large-scale departure from homogeneity, uniformity, or isotropy. Starting from a grossly homogeneous earth, with uniformly distributed isotropic heat sources, either radiogenic or gravitational, how does one go about producing such ordered structures as the magnetic field?

The answer to this question is, of course, convection. Convection is a prime example of a nonequilibrium state in which highly disorganized heat sources and random fluctuations somehow contrive to produce coherent, organized flow and an ordered thermal structure (Prigogine, 1978). That convection is a major geophysical process can no longer be doubted. No theory of the earth based on transfer of heat only by conduction has been able to account satisfactorily for the production of basaltic magma, which follows almost automatically from the convection hypothesis (Verhoogen, 1954). Convection is the main heat carrier in the mantle, and the prime mover of plates.

But there can be no natural convection without gravity. Gravity and, in the core, rotation of the earth are the factors through which structures develop. Perhaps it is not surprising that so much of this book should be concerned with interactions of thermal and gravitational fields.

References

Acheson, D. J., and R. Hide, Hydromagnetics of rotating fluids, *Rep. Prog. Phys.*, *36*, 159–221, 1973.

Ahrens, T. J., Petrological properties of the upper 670 km of the earth's mantle; geophysical implications, *Phys. Earth Planet. Inter.*, *7*, 167–186, 1973.

Akimoto, S., The system MgO-FeO-SiO_2 at high pressures and temperatures. Phase equilibria and elastic properties, *Tectonophysics*, *13*, 161–187, 1972.

Albarede, F., The heat flow–heat generation relationship: an interaction model of fluids with cooling intrusions, *Earth Planet. Sci. Lett.*, *27*, 73–78, 1975.

Alder, B. J., Is the mantle soluble in the core?, *J. Geophys. Res.*, *71*, 4973–4979, 1966.

Alder, B. J., and M. Trigueros, Suggestion of a eutectic region between the liquid and solid core of the earth, *J. Geophys. Res.*, *82*, 2535–2539, 1977.

Anderson, P. D., and R. Hultgren, The thermodynamics of solid iron at elevated temperatures, *Trans. Metall. Soc. AIME*, *224*, 842–845, 1962.

Backus, G. E., Gross thermodynamics of heat engines in deep interior of earth, *Proc. Nat. Acad. Sci. USA*, *72*, 1555–1558, 1975.

Birch, F., Elasticity and constitution of the earth's interior, *J. Geophys. Res.*, *57*, 227–286, 1952.

Birch, F., Some geophysical applications of high-pressure research, in *Solids Under Pressure*, ed. by W. Paul and D.M. Warschauer, McGraw-Hill, New York, 1963.

Birch, F., Speculations on the earth's thermal history, *Geol. Soc. Am. Bull.*, *76*, 133–154, 1965a.

Birch, F., Energetics of core formation, *J. Geophys. Res.*, *70*, 6217–6221, 1965b.

Birch, F., The melting relations of iron, and temperatures in the earth's core, *Geophys. J. R. Astron. Soc.*, *29*, 373–387, 1972.

Bolt, B. A., The density distribution near the base of the mantle and near the earth's center, *Phys. Earth Planet. Inter.*, *5*, 301–311, 1972.

Bolt, B. A., and R. Uhrhammer, Resolution techniques for density and heterogeneity in the earth, *Geophys. J. R. Astron. Soc.*, *42*, 419–435, 1975.

Boschi, E., Melting of iron, *Geophys. J. R. Astron. Soc.*, *38*, 322–334, 1974.

Boschi, E., The melting relations of iron and temperatures in the earth's core, *Riv. Nuovo Cimento*, Series 2, *5*, 501–531, 1975.

Boyd, F. R., and P. H. Nixon, Structure of the upper mantle under Lesotho, *Carnegie Inst. Washington Yearb.*, *72*, 431–445, 1973.

Braginsky, S. I., Structure of the F layer and reasons for convection in the earth's core, *Dokl. Akad. Nauk SSSR*, *149*, 8–10, 1963.

Braginsky, S. I., Magnetohydrodynamics of the earth's core, *Geomagn. Aeron.*, *4*, 698–712, 1964, (Engl. transl.).

Braginsky, S. I., Theory of the hydromagnetic dynamo, *Sov. Phys. JETP*, *20*, 1462–1471, 1965.

Braginsky, S. I., On the nearly axially-symmetrical model of the hydromagnetic dynamo of the earth, *Phys. Earth Planet. Inter.*, *11*, 191–199, 1976.

Brett, R., The current status of speculations on the composition of the core of the earth, *Rev. Geophys. Space Phys.*, *14*, 375–383, 1976.

Brett, R., and P. M. Bell, Melting relations in the Fe-rich portion of the system Fe-FeS at 30 kbar pressure, *Earth Planet. Sci. Lett.*, *6*, 479–482, 1969.

Bukowinski, M. S. T., The effect of pressure on the physics and chemistry of potassium, *Geophys. Res. Lett.*, *3*, 491–494, 1976.

Bukowinski, M. S. T., A theoretical equation of state for the inner core, *Phys. Earth Planet. Inter.*, *14*, 333–344, 1977.

Bullard, E. C., and D. Gubbins, Generation of magnetic fields by fluid motions of global scale, *Geophys. Astrophys. Fluid Dyn.*, *8*, 43–56, 1977.

Bullen, K. E., The variation of density and the ellipticities of strata of equal density within the earth, *Mon. Not. R. Astron. Soc., Geophys. Suppl.*, *3*, 395–401, 1936.

Bullen, K. E., The problem of the earth's density variation, *Bull. Seismol. Soc. Am.*, *30*, 235–250, 1940.

Bullen, K. E., An earth model based on a compressibility-pressure hypothesis, *Mon. Not. R. Astron. Soc., Geophys. Suppl.*, *6*, 50–59, 1950.

Bullen, K. E., Cores of the terrestrial planets, *Nature*, *243*, 68, 1973a.

Bullen, K. E., On planetary cores, *Moon*, *7*, 384, 1973b.

Busse, F. H., On the stability of two-dimensional convection in a layer heated from below, *J. Math. Phys.*, *46*, 140–150, 1967.

Busse, F. H., Patterns of convection in spherical shells, *J. Fluid Mech.*, *72*, 67–85, 1975a.

Busse, F. H., A model of the geodynamo, *Geophys. J. R. Astron. Soc.*, *42*, 437–459, 1975b.

Busse, F. H., and P. G. Cuong, Convection in rapidly rotating spherical fluid shells, *Geophys. Astrophys. Fluid Dyn.*, *8*, 17–44, 1977.

Calame, O., and J. D. Mulholland, Lunar tidal acceleration determined from laser range measurements, *Science*, *199*, 977–978, 1978.

Cannon, J. F., Behavior of the elements at high pressures, *J. Phys. Chem. Ref. Data*, *3*, 781–824, 1974.

Cathles, L. M., *The Viscosity of the Earth's Mantle*, Princeton University Press, Princeton, N.J., 1975.

Chandrasekhar, S., *Hydrodynamic and Hydromagnetic Stability*, Clarendon Press, Oxford, 1961.

Chapman, D. S., and H. N. Pollack, Global heat flow, a new look, *Earth Planet. Sci. Lett.*, *28*, 23–32, 1975.

Chapman, D. S., and H. N. Pollack, Regional geotherms and lithospheric thickness, *Geology*, *5*, 265–268, 1977.

Christensen, N. I., and M. H. Salisbury, Structure and consitution of the lower oceanic crust, *Rev. Geophys. Space Phys.*, *13*, 57–86, 1975.

Clark, S. P., Jr., and A. E. Ringwood, Density distribution and constitution of the mantle, *Rev. Geophys., 2*, 35–88, 1964.
Cleary, J. R., The D'' region, *Phys. Earth Planet. Inter., 9*, 13–27, 1974.
Craig, H., J. E. Lupton, J. A. Welhan, and R. Poreda, Helium isotope ratios in Yellowstone and Lassen Park volcanic gases, *Geophys. Res. Lett., 5*, 897–900, 1978.
Davies, G. F., Whole-mantle convection and plate tectonics, *Geophys. J. R. Astron. Soc., 49*, 459–486, 1977.
Dickman, S. R., Secular trend of the earth's rotation pole: Consideration of motion of the latitude observatories, *Geophys. J. R. Astron. Soc., 51*, 229–244, 1977.
Drabble, J. R., and H. J. Goldsmid, *Thermal Conduction in Semiconductors*, Pergamon Press, London, 1961.
Dubrovskii, V. A., and V. Pankov, On the composition of the earth's core, *Phys. Solid Earth*, 452–455, 1972, (Engl. transl.).
Dziewonski, A., A. L. Hales, and E. R. Lapwood, Parametrically simple earth models consistent with geophysical data, *Phys. Earth Planet. Inter., 10*, 12–48, 1975.
Eddington, A. S., *The Internal Constitution of the Stars*, Cambridge University Press, London, 1926.
Elsasser, W. M., P. Olson, and B. D. Marsh, The depth of mantle convection, *J. Geophys. Res., 84*, 147–155, 1979.
Filipov, S. I., N. B. Kazakov, and L. A. Pronin, Velocities of ultrasonic waves, compressibility of liquid metals, and their relation to various physical properties (in Russian), *Izv. Vyssh. Uchebn., Zaved. Chern. Metall., 9*, 8–14, 1966.
Flasar, F. M., and F. Birch, Energetics of core formation, a correction, *J. Geophys. Res., 78*, 6101–6103, 1973.
Foster, T. D., Intermittent convection, *Geophys. Fluid Dyn., 2*, 201–217, 1971.
Ganapathy, R., and E. Anders, Bulk composition of the moon and earth, estimated from meteorites, *Geochim. Cosmochim. Acta*, Suppl. 5, 1181–1206, 1974.
Gilman, P. A., Nonlinear dynamics of Boussinesq convection in a deep rotating spherical shell, *Geophys. Astrophys. Fluid Dyn., 8*, 93–135, 1977.
Glansdorff, P., and I. Prigogine, *Thermodynamic Theory of Structure, Stability, and Fluctuations*, Interscience, New York, 1971.
Goettel, K. A., Partitioning of potassium between silicates and sulphide melts: experiments relevant to the earth's core, *Phys. Earth Planet. Inter., 6*, 161–166, 1972.
Goettel, K. A., Models for the origin and composition of the earth, and the hypothesis of potassium in the earth's core, *Geophys. Surv., 2*, 369–397, 1976.
Graham, E. K., Elasticity and composition of the upper mantle, *Geophys. J. R. Astron. Soc., 20*, 285–302, 1970.
Graham, E. K., and D. Dobrzykowski, Temperatures in the mantle as inferred from simple compositional models, *Am. Mineral., 61*, 549–559, 1976.
Green, D. H., J. W. Morgan, and K. S. Heier, Thorium, uranium, and potassium abundances in peridotite inclusions and their host basalts, *Earth Planet. Sci. Lett., 4*, 155–166, 1968.
Gubbins, D., Can the earth's magnetic field be sustained by core oscillations?, *Geophys. Res. Lett., 2*, 409–412, 1975.
Gubbins, D., Observational constraints on the generation process of the earth's magnetic field, *Geophys. J. R. Astron. Soc., 47*, 19–39, 1976.
Gubbins, D., Energetics of the earth's core, *J. Geophys. (Z. Geofys.), 43*, 453–464, 1977.
Guggenheim, E. A., *Modern Thermodynamics by the Methods of Willard Gibbs*, Methuen, London, 1933.
Hanks, T. C., and D. L. Anderson, The early thermal history of the earth, *Phys. Earth Planet. Inter., 2*, 19–29, 1969.

Harte, B., Kimberlite nodules, upper mantle petrology and geotherms, *Phil. Trans. R. Soc. London, Ser. A, 288*, 487–500, 1978.

Heier, K. S., The distribution and redistribution of heat-producing elements in the continents, *Phil. Trans. R. Soc. London, Ser. A, 288*, 393–400, 1978.

Heier, K. S., and G. Grønlie, Heat flow–heat generation studies in Norway, in *Energetics of Geological Processes*, ed. by S. K. Saxena and S. Bhattacharji, pp. 217–235, Springer, New York, 1977.

Hewitt, J. M., D. P. McKenzie, and N. O. Weiss, Dissipative heating in convective flows, *J. Fluid Mech., 68*, 721–738, 1975.

Hide, R., The hydrodynamics of the earth's core, in *Physics and Chemistry of the Earth*, ed. by L. H. Ahrens, K. Rankama, and S. K. Runcorn, vol. 1, pp. 94–137, 1956.

Higgins, G. H., and G. C. Kennedy, The adiabatic gradient and the melting point gradient in the core of the earth, *J. Geophys. Res., 76*, 1870–1878, 1971.

Horai, K. I., and G. Simmons, An empirical relationship between thermal conductivity and Debye temperature for silicates, *J. Geophys. Res., 75*, 978–982, 1970.

Howard, L. N., Convection at high Rayleigh numbers, in *Proc. 11th Int. Congr. Appl. Mech.*, Munich, 1964, ed. by H. Görtler, Springer, Berlin, 1109–1115, 1966.

Hsui, A. T., D. L. Turcotte, and K. E. Torrance, Finite-amplitude thermal convection within a self gravitating fluid sphere, *Geophys. Fluid Dyn., 3*, 35–44, 1972.

Irving, E., and J. K. Park, Hairpins and superintervals, *Can. J. Earth Sci., 9*, 1318–1324, 1972.

Jacobs, J. A., *The Earth's Core*, Academic, New York, 1975.

Johnson, L. R., Array measurements of P velocities in the lower mantle, *Bull. Seismol. Soc. Am., 59*, 973–1008, 1969.

Jones, G. M., Thermal interaction of the core and the mantle and long-term behavior of the geomagnetic field, *J. Geophys. Res., 82*, 1703–1709, 1977.

Jordan, T. H., and D. L. Anderson, Earth structure from free oscillations and travel times, *Geophys. J. R. Astron. Soc., 36*, 411–459, 1974.

Kieffer, S. W., Lattice thermal conductivity within the earth and considerations of a relationship between the pressure dependence of the thermal diffusivity and the volume dependence of the Grüneisen parameter, *J. Geophys. Res., 81*, 3025–3030, 1976.

Kirshenbaum, A. D., and J. A. Cahill, The density of liquid iron from the melting point to 2500°K, *Trans. Metall. Soc. AIME, 224*, 816–819, 1962.

Knopoff L., and J. N. Shapiro, Comments on the interrelationships between Grüneisen's parameter and shock and isothermal equations of state, *J. Geophys. Res., 74*, 1439–1450, 1969.

Kraut, E. A., and G. C. Kennedy, New melting law at high pressures, *Phys. Rev., 151*, 668–675, 1966.

Krishnamurti, R., On the transition to turbulent convection, Part 2. The transition to time-dependent flow, *J. Fluid Mech., 42*, 309–320, 1970.

Kullerud, G., Sulfide phase relations, *Mineral. Soc. Am. Spec. Publ., 3*, 199–210, 1970.

Kurz, W., and B. Lux, Die schallengeschwindigkeit von eisen und eisenlegierungen im festen und flussigen zustand, *High Temp. High Pressures, 1*, 387–399, 1969.

Lachenbruch, A. H., Preliminary geothermal model for the Sierra Nevada, *J. Geophys. Res., 73*, 6977–6989, 1968.

Lambeck, K., Effect of tidal dissipation in the oceans on the moon's orbit and the earth's rotation, *J. Geophys. Res., 80*, 2917–2925, 1975.

Larimer, J. W., Composition of the earth: chondritic or achondritic?, *Geochim. Cosmochim. Acta, 35*, 769–786, 1971.

Lawson, A. W., On the high temperature heat conductivity of insulators, *J. Phys. Chem. Solids, 3*, 154–161, 1957.

Leppaluoto, D. A., Iron at pressures of the earth's core: properties inferred from the significant structure theory of liquids, Ph.D. thesis, Univ. of Calif. at Berkeley, 1972a.
Leppaluoto, D. A., Melting of iron by significant structure theory, *Phys. Earth Planet. Inter.*, 6, 175–181, 1972b.
Levy, E. H., Kinematic reversal schemes for the geomagnetic dipole, *Astrophys. J.*, 171, 635–642, 1972.
Lewis, J. S., Consequences of the presence of sulphur in the core of the earth, *Earth Planet. Sci. Lett.*, 11, 130–134, 1971.
Lewis, J. S., Metal/silicate fractionation in the solar system, *Earth Planet. Sci. Lett.*, 15, 286–290, 1972.
Liu, L. G., On the (γ, ϵ, l,) triple point of iron and the earth's core, *Geophys. J. R. Astron. Soc.*, 43, 697–705, 1975a.
Liu, L. G., Post-oxide phases of olivine and pyroxene and mineralogy of the mantle, *Nature*, 258, 510–512, 1975b.
Liu, L. G., Mineralogy and chemistry of the earth's mantle, *Geophys. J. R. Astron. Soc.*, 48, 53–62, 1977.
Liu, L. G., and W. A. Basset, The melting of iron up to 200 kbar, *J. Geophys. Res.*, 80, 3777–3782, 1975.
Loper, D. E., Torque balance and energy budget for the precessionally driven dynamo, *Phys. Earth Planet. Inter.*, 11, 43–60, 1975.
Loper, D. E., The gravitationally powered dynamo, *Geophys. J. R. Astron. Soc.*, 54, 389–404, 1978.
Loper, D. E., and P. H. Roberts, On the motion of an iron-alloy core containing a slurry, *Geophys. Astrophys. Fluid Dyn.*, 9, 289–321, 1978.
MacDonald, G. J. F., Geophysical deductions from observations of heat flow, chap. 7 in *Terrestrial Heat Flow*, ed. by W. H. K. Lee, Geophys. Monogr. No. 8, Am. Geophys. Union, Washington, D.C., 1965.
McKenzie, D. P., Some remarks on heat flow and gravity anomalies, *J. Geophys. Res.*, 72, 6261–6273, 1967.
McKenzie, D. P., The influence of the boundary conditions and rotation convection in the earth's mantle, *Geophys. J. R. Astron. Soc.*, 15, 457–500, 1968.
McKenzie, D. P., J. M. Roberts, and N. O. Weiss, Convection in the earth's mantle: towards a numerical simulation, *J. Fluid Mech.*, 62, 465–538, 1974.
McKenzie, D., and N. Weiss, Speculations on the thermal and tectonic history of the earth, *Geophys. J. R. Astron. Soc.*, 42, 131–174, 1975.
Malkus, W. V. R., Precessional torques as the cause of geomagnetism, *J. Geophys. Res.*, 68, 2871–2886, 1963.
Mao, H. K., Thermal and electrical properties of the earth's mantle, *Carnegie Inst. Washington Yearb.*, 72, 557–564, 1973a.
Mao, H. K., Observation of optical absorption and electrical conductivity in magnesiowüstite at high pressures, *Carnegie Inst. Washington Yearb.*, 72, 554–557, 1973b.
Mao, H. K., and P. M. Bell, Electrical conductivity and the red shift of absorption in olivine and spinel at high pressure, *Science*, 176, 403–406, 1972.
Mao, H. K., and P. M. Bell, Disproportionation equilibrium in iron-bearing systems at pressures above 100 kbar with application to chemistry of the earth's mantle, in *Energetics of Geological Processes*, ed. by S. K. Saxena and S. Bhattacharji, pp. 236–249, Springer, New York, 1977.
Mason, B., Composition of the earth, *Nature*, 211, 616–618, 1966.
Metchnik, V. J., M. T. Gladwin, and F. D. Stacey, Core convection as a power source

for the geomagnetic dynamo—a thermodynamic argument, *J. Geomagn. Geoelectr.*, 26, 405–415, 1974.

Monin, A. S., On some global problems of geophysical fluid dynamics (a review), *Proc. Nat. Acad. Sci. USA*, 75, 34–39, 1978.

Moore, D. R., and N. O. Weiss, Two-dimensional Rayleigh-Benard convection, *J. Fluid Mech.*, 58, 289–312, 1973.

Morgan, P., D. D. Blackwell, R. E. Spafford, and R. B. Smith, Heat flow measurements in Yellowstone Lake and the thermal structure of the Yellowstone Caldera, *J. Geophys. Res.*, 82, 3719–3732, 1977.

Murthy, V. Rama, Composition of the core and the early chemical history of the earth, in *The Early History of the Earth*, ed. by B. F. Windley, pp. 21–31, John Wiley, London, 1976.

Murthy, V. Rama, and H. T. Hall, The chemical composition of the earth's core: possibility of sulphur in the core, *Phys. Earth Planet. Inter.*, 2, 276–282, 1970.

Murthy, V. Rama, and H. T. Holl, The origin and composition of the earth's core, *Phys. Earth Planet. Inter.*, 6, 123–130, 1972.

Mysen, B. O., and A. L. Boettcher, Melting of a hydrous mantle: I. Phase relations of natural peridotite at high pressures and temperatures with controlled activities of water, carbon dioxide, and hydrogen, *J. Petrol.*, 16, 520–548, 1975.

Newton, R. R., Experimental evidence for a secular decrease in the gravitational constant G, *J. Geophys. Res.*, 73, 3765–3771, 1968.

Olson, P., Internal waves in the earth's core, *Geophys. J. R. Astron. Soc.*, 51, 183–215, 1977.

Oversby, V. M., and A. E. Ringwood, Potassium distribution between metal and silicate and its bearing on the occurrence of potassium in the core, *Earth Planet. Sci. Lett.*, 14, 345–347, 1972.

Palm, E., Nonlinear thermal convection, *Annu. Rev. Fluid Mech.*, 7, 39–61, 1975.

Parker, R. L., and D. W. Oldenburg, Thermal models of ocean ridges, *Nature*, 242, 137–139, 1973.

Peltier, W. R., Glacial-isostatic adjustment: II. The inverse problem, *Geophys. J. R. Astron. Soc.*, 46, 605–646, 1976.

Powell, R., The thermodynamics of pyroxene geotherms, *Phil. Trans. R. Soc. London, Ser. A*, 288, 457–469, 1978.

Prigogine, I., Time, structure, and fluctuations, *Science*, 201, 777–785, 1978.

Prigogine, I., and R. Defay, *Chemical Thermodynamics*, Longmans Green, London, 1954.

Richter, F. M., Finite amplitude convection through a phase boundary, *Geophys. J. R. Astron. Soc.*, 35, 267–276, 1973.

Ringwood, A. E., Composition and origin of the earth, *Publ. 1299, Res. Sch. Earth Sci.*, Aust. Nat. Univ., Canberra, 1977.

Ringwood, A. E., and A. Major, The system Mg_2SiO_4-Fe_2SiO_4 at high pressures and temperatures, *Phys. Earth. Planet. Inter.*, 3, 89–108, 1970.

Rochester, M. G., The earth's rotation, *Eos*, 54, 769–780, 1973.

Rochester, M. G., J. A. Jacobs, D. E. Smylie, and K. F. Chong, Can precession power the geomagnetic dynamo?, *Geophys. J. R. Astron. Soc.*, 43, 661–678, 1975.

Ross, J. E., and L. H. Aller, The chemical composition of the sun, *Nature*, 191, 1223–1229, 1976.

Roy, R. F., D. D. Blackwell, and E. R. Decker, Continental heat flow, in *The Nature of the Solid Earth*, ed. by E. C. Robertson, pp. 506–543, McGraw-Hill, New York, 1972.

Rybach, L., D. Werner, S. Mueller, and G. Berset, Heat flow, heat production and crustal dynamics in the central Alps, Switzerland, *Tectonophysics*, 41, 113–126, 1977.

Safronov, V. S., *Evolution of the Protoplanetary Cloud and Formation of the Earth and Planets*, Nauka, Moscow (NASA-TTF-677), 1969.
Schatz, J. F., and G. Simmons, Thermal conductivity of earth materials at high temperatures, *J. Geophys. Res., 77,* 6966–6983, 1972.
Schubert, G., and D. L. Turcotte, Phase changes and mantle convection, *J. Geophys. Res., 76,* 1424–1432, 1971.
Schubert, G., D. A. Yuen, and D. L. Turcotte, Role of phase transitions in a dynamic mantle, *Geophys. J. R. Astron. Soc., 42,* 705–735, 1975.
Sclater, J. G., and J. Francheteau, The implication of terrestrial heat flow observations on current tectonic and geochemical models of the crust and upper mantle of the earth, *Geophys. J. R. Astron. Soc., 20,* 509–542, 1970.
Sharpe, H. N., and W. R. Peltier, Parameterized mantle convection and the earth's thermal history, *Geophys. Res. Lett., 5,* 737–740, 1978.
Solomon, S. S., Geophysical constraints on radial and lateral temperature variations in the upper mantle, *Am. Mineral., 61,* 788–803, 1976.
Somerville, M., Elasticity, Constitution and temperature of earth's lower mantle, Ph.D. thesis, Univ. of Calif. at Berkeley, 1977.
Stacey, F. D., A thermal model of the earth, *Phys. Earth Planet. Inter., 15, 341–348,* 1977.
Sterrett, K. F., W. Klement, and G. C. Kennedy, The effect of pressure on the melting of iron, *J. Geophys. Res., 70,* 1979–1984, 1965.
Stewart, R. M., Composition and temperature of the outer core, *J. Geophys. Res., 78,* 2586–2597, 1973.
Thomsen, L., Equations of state and the interior of the earth, in *Mantle and Core in Planetary Physics,* ed. by J. Coulomb and M. Caputo, Proc. Int. Sch. Phys. Enrico Fermi, vol. 50, Academic, New York, 1971.
Tozer, D. C., The present thermal state of the terrestrial planets, *Phys. Earth Planet. Inter., 6,* 182–197, 1972.
Tozer, D. C., The thermal state and evolution of the earth and terrestrial planets, *Sci. Prog., Oxford., 64,* 1–28, 1977.
Tuerpe, D. R., and R. N. Keeler, Anomalous melting transition in the significant structure theory of liquids, *J. Chem. Phys., 47,* 4283–4285, 1967.
Turcotte, D. L., and E. R. Oxburgh, Finite amplitude convection cells and continental drift, *J. Fluid Mech., 28,* 29–42, 1967.
Usselman, T. M., Experimental approach to the state of the core: Part I. The liquidus relations of the Fe-rich portion of the Fe-Ni-S system from 30 to 100 kbar, *Am. J. Sci., 275,* 278–290, 1975a.
Usselman, T. M., Experimental approach to the state of the core: Part II. Composition and thermal regime, *Am. J. Sci., 275,* 291–303, 1975b.
Verhoogen, J., Von Zeipel's theorem and convection in the earth, *Trans. Am. Geophys. Union, 29,* 361–365, 1948.
Verhoogen, J., The adiabatic gradient in the mantle, *Trans. Am. Geophys. Union, 32,* 41–43, 1951.
Verhoogen, J., Petrological evidence on temperature distribution in the mantle of the earth, *Trans. Am. Geophys. Union, 35,* 85–92, 1954.
Verhoogen, J., Heat balance of the earth's core, *Geophys. J. R. Astron. Soc., 4,* 276–281, 1961.
Verhoogen, J., Phase changes and convection in the earth's mantle, *Phil. Trans. R. Soc. London, Ser. A, 258,* 276–283, 1965.
Verhoogen, J., Thermal regime of the earth's core, *Phys. Earth Planet. Inter., 7,* 47–58, 1973.

Walzer, V., Convection currents in the earth's mantle and the spherical harmonic development of the topography of the earth, *Pure Appl. Geophys.*, *87*, 73–92, 1971.

Wang, C. Y., Temperature in the lower mantle, *Geophys. J. R. Astron. Soc.*, *27*, 29–36, 1972.

Watt, J. P., and R. J. O'Connell, Mixed-oxide and perovskite-structure model mantles from 700–1200 km, *Geophys. J. R. Astron. Soc.*, *54*, 601–630, 1978.

Wetherill, G. W., Interplanetary bodies and solar system history, *Carnegie Inst. Washington Yearb.*, *75*, 151–169, 1976.

Wetherill, G. W., Evolution of small bodies in the solar system, *Carnegie Inst. Washington Yearb.*, *76*, 761–792, 1977.

Williams, D. L., and R. P. von Herzen, Heat loss from the earth: new estimate, *Geology*, *2*, 327–328, 1974.

Wyllie, P. J., *The Dynamic Earth: Textbook in Geosciences*, John Wiley, New York, 1971.

Yukutake, T., The effect of change in the geomagnetic dipole moment on the rate of the earth's rotation, *J. Geomagn. Geoelectr.*, *24*, 19–47, 1972.

Author Index

Acheson, D.J., 73
Ahrens, T.J., 43
Akimoto, S., 35, 36, 43
Albarede, F., 26
Alder, B.J., 49, 57
Anders, E., 23, 24
Anderson, D.L., 16, 17, 43, 123

Backus, G.E., 76, 85
Bell, P.M., 47, 51, 60
Birch, F., 19, 20, 38, 42, 49, 58
Boettcher, A.L., 32
Bolt, B.A., 37, 54, 63
Boschi, E., 56–60
Boyd, F.R., 33
Braginsky, S.I., 21, 67, 73, 77
Brett, R., 50, 60
Bukowinski, M.S.T., 25, 53, 55, 64, 65, 66, 73, 74
Bullen, K.E., 35, 36
Busse, F., 73, 106, 107, 111

Cahill, J.A., 55
Calame, O., 22
Cannon, J.F., 58
Cathles, L.M., 41
Chandrasekhar, S., 73, 105
Chapman, D.S., 3, 9, 31

Christensen, N.I., 4
Clark, S.P., Jr., 29
Cleary, J.R., 37
Corcos, G.M., 107
Craig, H., 24
Cuong, P.G., 111

Davies, G.F., 114
Defay, R., 62
Dickman, S.R., 12
Dobrzykowski, D., 39, 41
Dubrovskii, V.A., 49
Dziewonski, A.A., 53, 54

Eddington, A.S., 113
Elsasser, W.M., 45, 110

Filipov, S.I., 55
Flasar, F.M., 20
Foster, T.D., 110–111
Francheteau, J., 3

Ganapathy, R., 23, 24
Gilman, P.A., 111
Glansdorff, P., 81
Goettel, K.A., 25
Graham, E.K., 36, 39, 41
Green, D.H., 23

Author Index

Grønlie, G., 28
Gubbins, D., 21, 67, 71, 73, 88, 90, 91, 95
Guggenheim, E.A., 69

Hall, H.T., 50, 51
Hanks, T.C., 16, 17, 123
Harte, B., 34
Heier, K.S., 28
Hewitt, J.M., 80, 82, 88, 103, 117
Hide, R., 73, 79
Higgins, G.H., 68
Horai, K.I., 45, 46, 47
Howard, L.N., 110
Hsui, A.T., 106

Irving, E., 120

Jacobs, J.A., 56
Johnson, L.R., 37
Jones, G.M., 44, 45, 48, 107, 110, 120
Jordan, T.H., 43

Keeler, R.N., 59
Kennedy, G.C., 57–59, 68
Kieffer, S.W., 46, 47
Kirshenbaum, A.D., 55
Knopoff, L., 55
Kraut, E.A., 56–58
Krishnamurti, R., 107, 108
Kullerod, G., 62
Kurz, W., 55

Lachenbruch, A.H., 26
Lambeck, K., 22
Larimer, J.W., 23, 24
Lawson, A.W., 46
Leppaluoto, D.A., 48, 59, 60, 92, 93, 96
Lewis, J.S., 24
Liu, L.G., 36, 37, 53, 58
Loper, D.E., 21, 67, 88, 89, 94, 98
Lux, B., 55

MacDonald, G.J.F., 29, 30
Major, A., 35, 36
Mao, H.K., 46, 47, 48, 50
Mason, B., 50
McKenzie, D., 3, 108, 109, 110, 111, 113, 118, 119
Metchnik, V.J., 77
Monin, A.S., 20, 21
Moore, D.R., 48

Morgan, P., 4
Mulholland, J.D., 22
Murthy, V.R., 50, 51, 52
Mysen, B.O., 32

Newton, R.R., 22
Nixon, P.H., 33

O'Connell, R.J., 39
Oldenburg, D.W., 3
Olson, P., 68–69
Oversby, V.M., 25
Oxburgh, E.R., 41, 47

Palm, E., 112
Pankov, V., 51
Park, J.K., 120
Parker, R.L., 3
Peltier, W.R., 18, 41
Pollack, H.N., 3, 9, 31
Powell, R., 34
Prigogine, I., 62, 81, 124

Richter, F., 109, 114, 118
Ringwood, A.E., 25, 29, 35, 36, 51
Roberts, P.H., 67
Rochester, M.G., 21, 67
Roy, R.F., 26, 27
Rybach, L., 28

Safronov, V.S., 17
Salisbury, M.H., 4
Schatz, J.F., 47
Schubert, G., 114–118
Sclater, J.G., 3
Shapiro, J.N., 55
Sharpe, H.N., 18
Simmons, G., 45, 46, 47
Solomon, S.S., 31
Somerville, M., 42, 43, 44
Stacey, F.D., 55, 61, 62, 63, 94
Sterrett, K.F., 57–59
Stewart, R.M., 52, 53

Thomsen, L., 31, 44
Tozer, D.C., 122
Tuerpe, D.R., 59
Turcotte, D.L., 41, 47, 114–117

Uhrhammer, R., 54
Usselman, T.M., 60–64

Von Herzen, R.P., 3

Walzer, V., 106
Wang, C.Y., 39, 40, 41, 44
Watt, J.P., 39
Weiss, N.O., 48, 110, 118

Wetherill, G.W., 17
Williams, D.L., 3
Wyllie, P.J., 32

Yukutake, T., 22

Subject Index

Adams–Williamson equation, 38, 42, 52
Adiabatic gradient, 38, 39
 in core, 54–56, 68
 in divariant zone, 115–117
 in lower mantle, 41–42
 in upper mantle, 66
Argon, and potassium in earth, 24–25

β-phase, 35–36
Boussinesq approximation, 80, 117
Brunt-Väisälä frequency, 68

Chandler wobble, 12
Chemical convection, 67
Convection,
 in core, 67, 70
 effect on temperature, 29–30, 122
 at high Rayleigh number, 105–106, 108
 in mantle, patterns of, 104–110
Core,
 adiabatic gradient in, 54–56
 composition of, 49–52
 cooling of, 97–98
 equation of state of, 52–54
 formation of, 19–21
 heat output of, 95–100, 103, 110, 119
 influence on mantle convection, 118–120
 stability of, 68–70
 temperature, 65–66
Coriolis force, 73

D'-layer, temperature in, 36–44
D"-layer, temperatures in, 36–44
 as a thermal boundary layer, 44
 thermal conductivity of, 44–46
Debye temperature, 42–43, 46, 57
Distributed heat sources,
 effect on convection, 104–110
 Rayleigh number for, 105

Efficiency,
 from entropy balance, 85–87
 of gravitational dynamo, 95
 of mantle, 100–104
 of thermal dynamo, 75–83, 88
Electrical conductivity, in core, 71
Energy,
 to drive plates, 8–11
 seismic, 7
 sinks, 1–13
 sources, 15–28
Entropy balance equation,
 in core, 83–88
 in mantle, 101–104

138 Subject Index

Entropy production, 83–87
  and efficiency of dynamo, 85
  in mantle, 101–104
Eyring's significant-structure theory, 59

Fe-FeS system,
  phase diagram, 61
  volumetric relations in, 91–95
Fe-O system, 49–51
Fourier's law, 2

Geomagnetic field,
  energy in, 70–74
  generation of, 74
Geotherms,
  in lower mantle, 36–49
  in mantle transition zones, 35–36
  from nodules in kimberlites, 33–34
  "oceanic", 29
  "shield", 29
Gravitational dynamo, 88–99
Gravitational energy, 18
  and formation of core, 19–21
  and formation of inner core, 89–91
Grueneisen's ratio, 42, 44, 46, 52, 54, 55, 61

Heat conduction,
  entropy production by, 83–86, 101
  Fourier's law of, 2
Heat of crystallization of iron, 70, 96, 98
Heat flow,
  from mantle, 26–28
  and radioactivity of crust, 26
  regional variation of, 3
  at the surface, 2–3, 100–103
Heat loss from earth, 13
Heat, metamorphic, 5–7
Heat output, of core,
  from gravitational dynamo, 95–99
  and temperature gradient in layer D", 44–46
  for thermal dynamo, 87–88
Heat, volcanic, 4
Hugoniot parameters, for core, 52–53

Immiscibility, 95
  in Fe-FeS system, 62
  in Fe-O system, 51
Inner core,
  gravitational energy of, 89

temperature in, 53, 66
temperature gradient in, 74
Iron, melting point of, 56–60

Kilauea, 4
Kimberlites, nodules in, 33–34
Kinetic energy, of rotation, 11–12
Krakatoa, 5

Lherzolites, radioactivity of, 23
Lindemann's theory of melting, 56–57, 61
Lorentz force, 73–75, 79
Low-velocity zone, 31–33

Mantle,
  composition of, 36
  convection patterns in, 104–111
  influence on core convection, 120–121
  phase transitions in, 35–36
  radioactive content of, 23–26
  radiogenic heat in, 100
  temperature in, 31–49, 65–66
  temperature gradient in, 31–49
Melting,
  in Fe-S system, 60–64
  of iron, 56–60
  in low-velocity zone, 32
Metamorphic heat, 5–7
Moon, acceleration of, 21

Nusselt number, in lower mantle, 47

Oceanic ridges,
  heat flow at, 3
  rate of lava production at, 4
Ohmic dissipation, 71, 74, 75, 87, 90
Olivine-spinel transformation, 35–36
Original heat, 1, 15–18

Partial molar volume, 91–94
Perfect solution, 91, 96
Phase changes,
  and convection, 114–118
  in upper mantle, 35–36
Plate tectonics, 8–11
Potassium,
  content of earth, 24

and cooling of earth, 121
in core, 24–25
"High" K-model, 121–123
"Low" K-model, 121–123
relation to sulfur, 24
Prandtl number, 107, 108

Radiogenic heat, 23–28, 70, 104
Rayleigh number, 47
 for distributed heat sources, 105
 effect on convection pattern, 105–106, 111
 relation to Nusselt number, 47
Reynolds magnetic number, 72
Rotation,
 effect on heat flux from core, 111
 kinetic energy of, 11–12

Silicon, in core, 49
Strain energy, 7, 103–104
Structure, 2, 13, 123–124
Sulfur,
 in core, 48–52
 effect on melting of iron, 60–64
 in meteorites, 51

Taylor number, 111
Temperature,
 of accretion, 16–17
 of condensation in nebula, 16
Temperature, in core,
 from equation of state, 52–54
 at inner-core boundary, 56–66
 summary, 65–66
Temperature, in crust, effect of reactions on, 6
Temperature gradient, effect on efficiency of dynamo, 83
Temperature, in mantle,
 from conduction equation, 29–30
 at core-mantle boundary, 48
 in low-velocity zone, 31–33
 in lower mantle, 36–49
 in mantle transition zones, 35–36
 from nodules in kimberlites, 33–34
 summary, 64–66
Thermal conductivity, in layer D″, 45–47
Thorium,
 heat generation, 25
 in mantle, 24
Tidal friction, 21–23
Transition zones in mantle, 35, 36

Uplift of mountains, 8
Uranium,
 heat generation by, 25
 in mantle, 23, 100

Viscous dissipation, in mantle, 101–102
Volcanic heat, 4
Volumetric relations in Fe-FeS system, 91–95
Von Zeipel instability, 112–114

Yellowstone, 4